D0077663

TEXAS-WOMAN'S UNIVERSITY
3 9351 004929487

S

DATE DUE

DEC 0 9 2006			
GAYLORD			PRINTED IN U.S.A.

Projection for the
Performing Arts

This book is dedicated to those who, in the latter part of this century have remained as enthusiastic about projection as their counterparts were in its earlier formative years, and in doing so have ensured the survival and development of the art. In particular to Mervyn Heard, Eddie Biddle and Hermann Sorger.

This is the second book in the **Live Performance Technology Series**

Series editor: John Offord,
Editor of *Lighting + Sound International* and Chief Executive of the Professional Lighting and Sound Association

PROFESSIONAL LIGHTING
AND SOUND ASSOCIATION

The first book in the series was *The Focal Guide to Safety in Live Performance.*

Future titles will cover all areas of entertainment and presentation technology, including:

• lighting design, control and effects
• performance sound
• rigging and trussing

Projection for the Performing Arts

Graham Walne

Focal Press
An imprint of Butterworth-Heinemann Ltd
Linacre House, Jordan Hill, Oxford OX2 8DP

℞ A member of the Reed Elsevier plc group

OXFORD   LONDON   BOSTON
MUNICH   NEW DELHI   SINGAPORE   SYDNEY
TOKYO   TORONTO   WELLINGTON

First published 1995

© Graham Walne 1995

All rights reserved. No part of this publication
may be reproduced in any material form (including
photocopying or storing in any medium by electronic
means and whether or not transiently or incidentally
to some other use of this publication) without the
written permission of the copyright holder except
in accordance with the provisions of the Copyright,
Designs and Patents Act 1988 or under the terms of a
licence issued by the Copyright Licensing Agency Ltd,
90 Tottenham Court Road, London, England W1P 9HE.
Applications for the copyright holder's written permission
to reproduce any part of this publication should be addressed
to the publishers

British Library Cataloguing in Publication Data

Projection for the Performing Arts
  I. Walne, Graham
  790.2028

ISBN 0 240 51390 8

Library of Congress Cataloguing-in-Publication Data

Walne, Graham.
  Projection for the performing arts/Graham Walne.
  p.   cm.
  Includes index.
  ISBN 0 240 51390 8
  1. Projected scenery. 2. Theaters—Stage-setting and scenery.
  3. Stage lighting. 4. Motion picture projection. I. Title.
  PN2091.S8P76   1995   94–24063
  792'.025—dc20   CIP

Composition by Scribe Design, Gillingham, Kent
Printed and bound in Great Britain by
Hartnolls Limited, Bodmin, Cornwall

Contents

Introduction

People have been experimenting with projection ever since someone noticed their own shadow cast by the light of the sun, fire or candle, and early entertainments used this technique. The early study of optics, triggered by an interest in astrology, provided the necessary components of the projection system which we still use today. In that sense, the projector has not changed for thousands of years, it has simply become more sophisticated because of the benefits accruing from the use of improved light sources and materials. Whilst the Ancient Greeks would be staggered by the sophistication of the latest Pani scene projector, once they saw inside it, they would recognize many elements.

In more recent times, the phrase 'we're going to do it with projection' has been sufficient to strike simultaneous excitement and terror into the hearts of theatre people. Excitement because projected scenery has a unique quality which, done properly, contributes its own dimension to the production; and terror because few people really understand how it works, but, nevertheless, they have been given the responsibility of dealing with it. Often projection is a supposed soft option, a potential solution to an effect which no other department can deliver.

Its use in earlier times relied heavily on invention, but today the plethora of available equipment and the need to reduce expensive stage time have combined to squeeze much experimentation out of the technique. Yesterday's projectionists took risks, today's have to calculate everything beforehand. Good projection depends on a balance between the two.

Basics

Projection involves the throwing of light through an object (sometimes inaccurately called the image) onto a surface creating a picture; it is this picture which should more correctly be called the image. The *Oxford English Dictionary* usefully defines 'image' as an 'optical appearance . . . produced by light . . . from an object . . . refracted through a lens'. The surface on which the image is seen does not necessarily have to be a purpose-built projection screen since any surface at all will receive light, but we only see objects because light is reflected from a surface into our eye and it will be appreciated that some surfaces reflect more light than others, purpose-built screens are specially designed to maximize this reflection. Screens can also be designed to reflect light in specific directions relative to the angle at which they are viewed, as we shall see later. Any translucent material will serve as a back-projection surface but only purpose-made back-projection screens will transmit light efficiently.

The image will appear brighter if the projection surface (front or rear projection) is not receiving light from anywhere else which would wash out the image and one of the key factors governing successful projection is the control over this other light. This can come from two main areas, ambient light and direct light. Ambient light is created through the general wash of illumination bouncing up onto the projection surface, it is produced by neighbouring lightsources many of which may not actually be visible from the projection surface itself. Typical examples are sunlight and general stage lighting and we will see later how their effects can be minimized. Direct light is easier to control by angling either the source or the projection surface away from each other. The nature of the surfaces adjacent to the projection surface can influence how light is reflected, shiny mirror-like surfaces reflect precise shapes without much loss of intensity. This is called specular reflection and a common occurrence in theatre shows is the reflection of the follow spot beam from the shiny stage floor up onto a backcloth or projection screen. Diffuse reflection comes from a surface which is dull or textured and the resultant reflection will have lost most of its intensity and shape.

The intensity of the projected image is also governed by the size of the original which is being projected – usually the size of the slide or the film – by the power of the lightsource in the projector, and by the ability of the lens to transmit light without significant loss. One reason why a 5000 watt projector produces a brighter image than a 250 watt projector is that the former will have a larger slide (typically 180 mm square) and thus a larger aperture through which the light can pass. A 250 watt projector would be more associated with a 35mm slide, this dimension being the width of the actual transparency through which the light will pass.

Another factor is the nature of the object to be projected. The intensity and quality of an image thrown from slide or film which contains little detail or rich colour but bold simple lines or text will be superior to the equivalent size of image thrown from a slide or film with small highly coloured elements. Thus whilst each type of projector has a natural size of resultant image with which it would be naturally associated, this is influenced by the nature of the original object which is to be projected, even small projectors can successfully project quite large images if the slide is simple, clear, bold and not highly coloured.

The projected image will appear distorted if the projection surface is not at right angles to the centre-line of the projection beam. The usual type of distortion is called keystone distortion and results from one part of the projection surface being closer to the projector than another. Typical instances are when the projector is mounted high up in a rear circle control room or projection booth, or on the side of the stage at an angle to the projection surface. If the projection surface cannot be angled (the bottom of most cinema screens is actually angled up to face the projection room) then the slide can be made so that distortion is present within it and thus cancels the projection system's natural reversal format, slide top to screen bottom, left-to-right. Some slide projectors can also accommodate special lenses which minimize keystone distortion.

In the context of entertainment, the term 'projection' would cover the moving images associated with the commercial cinema. However, as this form of projection would rarely form a central component of a stage production, it is not covered in the same detail here as other aspects such as scenic projection – the term associated with the projection of scenery and usually the responsibility of the set designer. Scenic projection can be delivered through conventional slides, shadows, overhead projection systems associated with lecture theatres, or through cut-out masks inserted into stage spotlights (known as 'gobos', or 'patterns' in the USA). Effects in stage presentations are also often projected and there is a wide range of motorized discs, wheels and filmstrips which simulate weather or fire for example and which are available for most proprietary projection systems. Finally laser projection, although not a frequent component of theatre presentations, is nevertheless an almost essential part of concerts by popular artists and the constant development of laser devices promises that they may become more than merely a tool to produce intense line drawings.

Why project?

In the context of a theatrical presentation, or a performance of any kind, be it concert, trade show or even lecture, there are various reasons why projections are chosen.

Scenic

The projection of an image which forms part or all of a set is usually known as 'scenic' projection and is usually the province of the set designer. Generally, but not exclusively, scenic projections are slides which are made up from the artwork drawn or chosen by the set designer. Alternatively scenic projection can be delivered through the creation of shadows, or from transparencies mounted on an overhead projector, or through the use of metal or glass slides inserted in the gate of profile spotlights (known as 'lekos' in the USA).

There are essentially three reasons for using scenic projection. First the quality of the projected image is unique and unlike an image realized through any other method such as lighting up a painted surface. Projected images can have an iridescence which a solid or opaque surface cannot; frequently the effect which the designer wants can only be achieved through projection as was the case with Timothy O'Brien and Tazeena Firth's design for the *Knot Garden* at the Royal Opera House (see later). In this case the spectacular and memorable result could not have been achieved by (say) paint. Perhaps paint would have made an interesting picture but it would have been totally different, and certainly not as magical, as that which the designers wanted. Projectors for scenic work tend to be large and powerful and located backstage, the involvement of this medium in the production requires careful planning as we shall see later.

Second, scenic projection can be a simple and cost effective way of creating an environment when the scenes are too short to justify a three-dimensional construction, and/or when the process of changing one three-dimensional scene for another is impractical. If the theatre lacks the flying system, wingspace, labour or motivepower associated with scene changes then it may be wiser to project and simply change the slide instead. It was the speed of the many scenes which led Richard Pilbrow and Robert Ornbo to project scenery in the revue *One over the Eight* as we shall see later in Chapter Four. This was an early demonstration of the use of this technique and led the way for others to follow.

Third, projection is also used to add texture to the set, for example the dappling effect of sunlight coming through the leaves of trees. In this case the projection will almost certainly be a gobo (see above) and whilst it may be suggested by the set designer or director, its realization will be the province of the lighting designer. There is a wide range of proprietary gobos covering (for example) a range of foliage and abstract shapes, but gobos can always be made to designers' own artwork – useful if the gobo is taking the place of a photographic slide.

Information

Projections are a useful way of conveying information to the audience – typical cases are montages of photographs, newsprint and text in anthologies of someone's life, or in concerts which contain songs which can usefully be supported by a visual image. In these cases there are likely to be many images during the evening, projected in quick succession and often side by side or overlapping. In theatre the projection of information often links scenes together and provides a distraction whilst the set is being changed elsewhere on the stage. Projections are also used in opera to show the translation of the libretto for the audience (see surtitles in Chapter Three).

Effects

The effect of rain, snow, fire and water, to name the main elements, can be very realistically projected from a wide range of tried and tested equipment from an equally wide range of manufacturers. In each case the projected effect is not the whole story because other elements are involved such as sound and mechanics. Projected effects require careful choreographing into the production otherwise there is a danger that they might distract the audience.

Thus projection is a particular route in the solution of particular problems, or it can be chosen consciously to add a different element to the visual appreciation of a performance.

1
The developments in scenic projection

Early Projection

In his excellent chronology of pre-cinema projection, *Dates and Sources*, Franz Paul Liesegang suggests that the earliest formal projection system was that developed by Athanasius Kircher who published his research in Rome in 1646 and again later in Amsterdam in 1671. Kircher's system was based on even earlier experiments of mirror writing, described by Johann Baptista Porta in 1589. Kircher's system relies on the reflection from a mirror incorporating the image of an object in front of, or painted on the mirror. Sunlight or candlelight was used and the image focused by means of a 'bi-convex' lens. Kircher's system was later used by Andreas Tacquet in either 1653 or 1654 to project pictures of a journey from China to the Netherlands which his colleague had taken. This appears to be the first public demonstration of a projected image. (Figure 1.1).

Liesegang continues to list a variety of devices which were developed throughout the remaining years of the seventeenth century and which were mostly called 'Laterna Magica' (a term later associated with Josef Svoboda – see below) this term became Magic Lantern in the UK and USA. Light sources continued to be either sunlight or candlelight; some projectors collected the light via a mirror and some had objective lenses but no condenser lens system. People became aware that the lantern could be useful for educational purposes although its use in this way was not widespread for some time. One supporter of this idea was Johann Zahn who

Figure 1.1 Tacquet's demonstration of Kircher's projection system, thought to be the first public demonstration 1653/4. *Source*: Leisegang, 1986, p. 10.

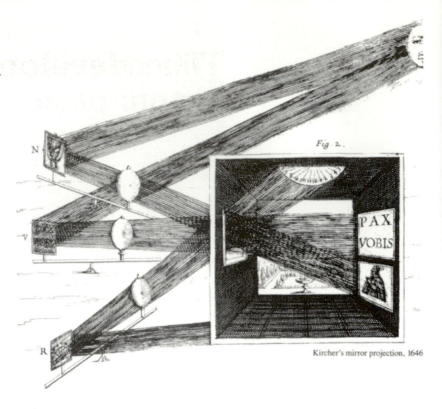

Kircher's mirror projection, 1646

in 1685/86 worked extensively with projections featuring small living animals which were contained in a trough, the image presumably being projected via a mirror system (Figure 1.2).

By the early years of the eighteenth century mechanical slides became available and the use of oil burners appears to have become more common, presumably because candlelight was not sufficiently powerful to match the increasing demands on the optical system. By the middle of the eighteenth century limelight was replacing both candles and oil, and the Argand lamp was common. One of Zahn's ideas developed into the solar microscope which was demonstrated at the Royal Society in London in 1739. A more sophisticated version was then developed by a London optician, who added a mirror to collect the sunlight, and Liesegang reports that sales were very popular. These devices projected opaque objects and so the term 'camera obscura' emerged at this time. The opaque objects increased in size until one system in Paris used real people! The eighteenth century was a lively time as the application of the lantern to medical and educational purposes encouraged experiments with optical systems and by the end of the century the double condenser lens system had been

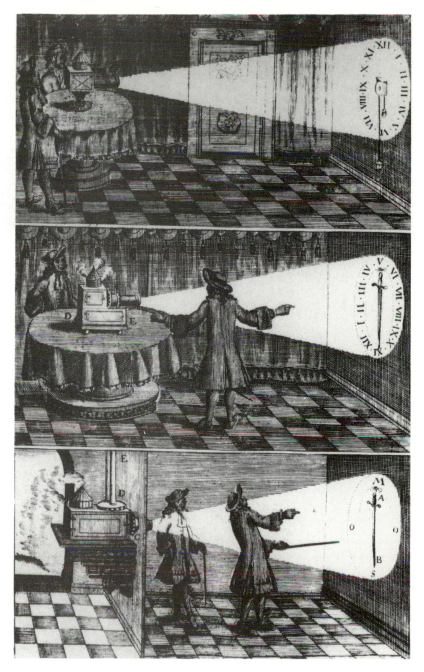

Figure 1.2 Zahn's Magic Lantern system which also used discs, 1685/6. *Source*: Leisegang, 1986, p. 13.

established. It was reported that this produced a clearer picture and one without 'spherical aberration' (see glossary and later) (Figure 1.3).

Magic lanterns are popularly associated with the term 'Phantasmagoria', which means ghostly apparitions, and which was used to describe lantern entertainments in the later

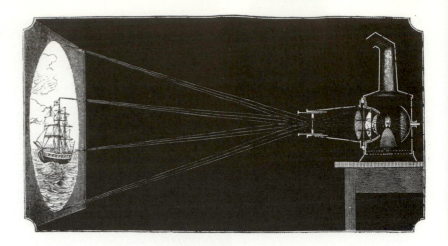

Figure 1.3 Magic Lantern with triple condenser lens. *Source*: Leisegang, 1986, p. 19.

eighteenth century; various shows in Paris, Vienna, St Petersburg and London were all described thus. The 'Phantasmagoria' was first introduced into the UK in 1802 by Paul de Philipsthal who exhibited at the Lyceum Theatre in the Strand. The advertisement for the event announced that 'Phantoms or apparitions of the dead or absent, in a way more compleatly illusive than has ever been offered to the eye in a public Theatre, as the objects freely originate in the air, and unfold themselves in various forms and sizes, such as Imagination alone has hitherto painted them'. It is worth noting that this is one of the first public occasions when the effects were back projected and thus the tricks were hidden (although this technique was used earlier for smaller and more private showings) (Figure 1.4).

Figure 1.4 Robertson's *Phantasmagoria* in 1845 showing the back projection system. *Source*: Leisegang, 1986, p. 18.

Figure 1.5 A Limelight showing the tubes feeding oxygen and hydrogen, the operator is adjusting the burning position of the piece of lime, this is a later model, earlier ones did not include lenses, (Science Museum). *Source*: HMSO, 'Lighting', p. 14.

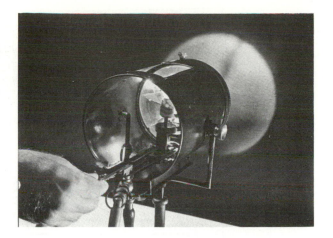

The development of limelight made larger and brighter projections possible. The limelight was developed by a Dr Goldworthy Gurney in 1822 and was used by Lieutenant Drummond in 1825 as a signal light during the geodesic survey in Ireland. Since this was (at that time) the brightest lightsource, it attracted much attention elsewhere and it quickly became the main source of light for magic lanterns as well as the limelights (early follow spots) in theatres (Figure 1.5).

There is some dispute about the earliest use of 'dissolving views' involving two or more projectors. According to Liesegang, the earliest use of 'dissolving views' appears to have been 1798 in Philipsthal's Phantasmagoria with the technique in the UK generally being thought to be the work of Henry Langdon Childe who demonstrated it in the 1830s. Typical of these effects was one which a French projectionist demonstrated in 1838, when a winter snow scene dissolved into a summer landscape. Childe also is credited with the invention of the Chromatrope, in which two patterns interact, the effect is produced by revolving two patterned discs in opposite directions; a modern version of this device is still in existence for 2 kW and 2.5 kW moving effects projectors produced in the UK. Childe's invention was recorded by J. H. Pepper later to become famous for the ghost projection which now bears his name (see below).

The development of electricity which, for a time, existed alongside coal gas and limelight as a lightsource, enabled the carbon arc to be developed as a lightsource for projectors. Initial systems required that the arcs were manually fed towards each other as they burned but, in 1849 and 1850, Foucault and Duboscq developed an automatic system of

Figure 1.6 Robertson's
demonstration of twin lanterns back
projected at the Royal Polytechnic
Institution. *Source*: Leisegang,
1986, p. 20.

feeding the arcs which made the arc a practical lightsource for projection since it was more constant. Duboscq also produced a lantern with interchangeable limelight or arc lightsource and its design became a standard for some years. Another notable date is 1870 when, according to the 'Optical Magic Lantern Journal' of 1893 an F. Bartlett became the first person to use a double slide carrier (Figure 1.6).

The USA was also not without its inventors. A Philadelphia optician L. J. Marcy replaced the oil wick with two paraffin wicks each in line with the axis of the projector, he called his machine the Sciopticon and its performance apparently boosted the popularity of the magic lantern, not only in the USA but also in Europe where the Sciopticon reached in 1873 and 1874. The name 'Sciopticon' was later used by other manufacturers of effects equipment.

In addition to the developments during the nineteenth century described above, various people in Europe were examining the eye's ability to 'remember' an image for a brief period after the image has been taken away. This persistence of vision is know as 'Palings Enigma' which had been identified early in the century although Liesegang notes that Ptolemy had examined it in his book 'Optics' which was written about AD 130. The Victorians were treated to a range of small devices which could be rotated to demonstrate the phenomenon and the devices were generally called 'Thaumatropes' (Figure 1.7).

Figure 1.7 An early 'Thaumatrope' using the persistence of vision phenomenon to overlap the two separate images on the revolving disc. *Source*: Leisegang, 1986, p. 24.

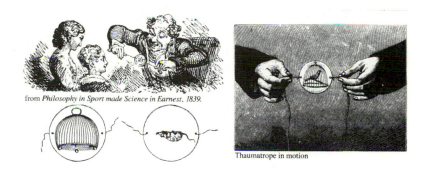

from *Philosophy in Sport made Science in Earnest, 1839.*

Thaumatrope in motion

Throughout the nineteenth century people experimented with more sophisticated devices to present the persistence of vision, and the devices included discs with sectors either cut out or coloured. The rate of revolution of the disc became known as the fusion frequency. Among people involved in this work was Helmholtz who also experimented with colour vision, thermodynamics and sound, and gave his name to a device for absorbing specific audio frequencies. Another person whose name is not generally associated with light is Michael Faraday, the discoverer of electrolysis in 1832. Faraday became professor of chemistry at the London Royal Institution in 1833 and became the first person to experiment with stroboscopic movement which he demonstrated with his 'Faraday disc' (Figure 1.8).

Figure 1.8 Phenakistoscope discs of 1833 (Science Museum). *Source*: Leisegang, 1986, p. 26.

Figure 1.9 The zoetrope, paper strips could be inserted inside the cylinder and when rotated the figures on the paper moved when viewed through the slits. *Source*: MOMI, p.7.

The 1830s were a rich period for the development of the moving image. Faraday's disc informed a number of other devices which consisted of cardboard discs containing slots and pictures and the devices went under a variety of names, 'phenakistoscope' and 'fantoscope' being the most common. Some of these devices can now be seen in London's Science Museum and the Museum of the Moving Image also in London. In his book, Liesegang, makes the point that many inventions were re-inventions from an earlier age. Typical was the 'Daedalum' which took its name from Daedalus who was reputed to have invented a disc-like device. The system was at this time using the name 'zoetrope' and it consisted of a single cylinder which had slots cut into it and contained on the inside the pictures which were to be viewed through the slots (Figure 1.9).

In 1849 Jo Plateau combined a number of discoveries into his machine which formed the basis for a 'projection-phenakistoscope', in which the persistence of vision was linked to a rotating shutter from which a slot was cut. The beginnings of projecting moving images were under way. The next few years were full of variations and developments on this theme, although almost all involved the viewer looking into the eyepieces of a container rather than watching a projected picture as we know it today – the emphasis behind these devices was in the movement itself not in the projection of that movement. However in 1845 Franz Uchatius constructed a projector fitted with an oil or limelight source which was able to project moving images. A. E. Dolbear projected moving pictures in 1877 by means of either sunlight or limelight which was focused through a disc onto slides, apparently pictures of

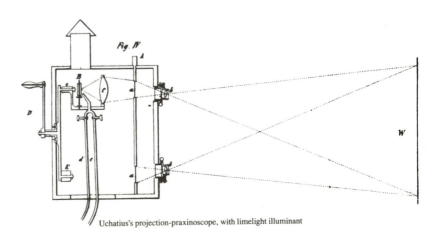

Uchatius's projection-praxinoscope, with limelight illuminant

Figure 1.10 Uchatius's projector. *Source*: Leisegang, 1986, p. 35.

3" (76 mm) wide were common and light levels quite accept-able at the time. For the next twenty years various patents were granted to a wide variety of devices, some claiming the ability to project. In 1853 Ludwig Dobler became the first to exhibit moving pictures using a device developed by Uchatius in which the lightsource (limelight) and the condenser revolved, the slides and the objective lenses remaining station-ary. This device is credited with being able to project satisfac-torily up to 2.5 m (Figure 1.10).

Magic Lantern

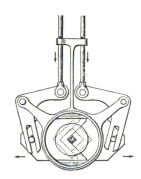

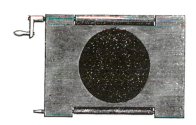

Figure 1.11 Magic Lantern slide with snow effect. *Source*: Rees, 1978, p. 82.

The Magic Lantern system involves the projection of slides from two or more projectors, each being synchronized in such a way that the beams of each projector completely overlap. Only one projector is operative at any one time, so the action of crossfading between projectors can create a series of overlapping or dissolving effects. Many slides also have moving parts operated by mechanical means within the slide mounting (Figure 1.11).

The Magic Lantern was very popular in the sixteenth and seventeenth centuries but it was not until the eighteenth century that some sophistication was applied to its presenta-tion. Previously the Lantern had been placed in the middle of the audience and thus much of the secret of the technique would have been lost. By removing the Lantern behind a screen, and back-projecting the effect, the mystery was maintained. The initial light source was oil but later limelight was used; the projectors were fitted with a variety of mechan-ical devices to adjust the iris and achieve some dimming effect, and zooming was achieved by mounting the projectors on tracks and moving them backwards and forwards in relation-ship to the screen (Figure 1.12).

The date of the first use of the Magic Lantern in the theatre is difficult to pin down; it is generally assumed to be 1827 in a production at the Adelphi Theatre in London of the *Flying Dutchman* in which the ghost ship was projected. However, there is also a view that Athanasius Kircher (see above) used the device towards the end of the seventeenth century. In the middle of the eighteenth century devices such as Robert Smith's 'ocular harpsichord' and the 'Chase electric cyclorama' were popular and could have influenced Adolph Appia, who was popularly assumed to have introduced slide projection into the theatre.

Figure 1.12 These two Lanterns were fitted with an early self-adjusting focus device which operated as they moved in relation to the screen. *Source*: Rees, 1978, p. 83.

Magic Lanterns still exist and are available through specialist collectors and projectionists. The fine engineering and hand-painted slides of the later models are of value in themselves (Figure 1.13).

Figure 1.13 Lanterns like this one made in London in 1890 have now become collectors' items with their brass and mahogany finishes. *Source*: MOMI, p. 6.

Figure 1.14 Frames from an 1896 moving back projection. *Source*: Rees, 1978, p. 92.

Laterna Magica and Joseph Svoboda

The Laterna Magica technique involves the use of slide and film projection and simultaneous live action and is generally used to describe the entertainment which Josef Svoboda developed for the Czech Pavilion at the Brussel's World Fair in 1958 (although this was not the first use of this technique – see below). The system remained in the repertoire of the Czech National Theatre for some years. An earlier version took place in a performance in Paris in 1896 of *La Biche au Bois* in which a Magic Lantern was used to project a large image, moving pictures were also projected from a Demeny-Gaumont Chronophotographie machine in which the light source was limelight. At one point there was a staged explosion and imps emerged from the screen and danced around the stage. This is a technique which Svoboda used in many productions and which was enhanced by the development of purpose-made back-projection screen material in strips. Svoboda hung the strips vertically so that the gaps were almost invisible and then smoke and dancers would appear through the gaps in the material onto which images were simultaneously being projected. This effect is still possible using the Rosco or Gerriets back-projection material (see later for information on back projection) (Figure 1.14).

Svoboda, who was born in Czechoslovakia in 1920, became the head designer at the Czech National Theatre in 1948. His work, notably in London in 1967 at the National Theatre and at the Royal Opera House, did much to enhance the reputation of scenic projection and prove its value. Significantly Svoboda often lights his own productions keeping the overall level down (and hence the spill onto the screens) by the use of his distinctive low-voltage beamlight backlights and similar steeply angled follow spots. Svoboda also experiments with screen materials and has gone on record as saying that grey is the best colour. It is reported that some of his screens are nets sprayed on the projection side with black velvet fibres through an electrostatic process so that the fibres are vertical (presumably at right angles to the screen) which therefore absorbs the spill whilst the main beam shines through the net. Svoboda also sprays mirrors with particles to control absorption and reflection. The details of these techniques remain with Svoboda but they illustrate that it pays to experiment (Figure 1.15).

Figure 1.15 Settings and projection by Josef Svoboda for *Die Frau ohne Schatten* at the Royal Opera House, London, 1967. A gauze is used midstage and the downstage steps are raised to reveal other scenes on the underside. *Source: Tabs*, 1967, Vol. 25, no. 3, p. 17.

Pepper's Ghost

John Henry Pepper was Director and Professor of Chemistry at the Royal Polytechnic Institution in London which was founded in 1838. Pepper was, however, much more famous for the ghost effect (previously mentioned) which bears his name and which he demonstrated at the Institution.

In this effect Pepper made use of the ability of light to be reflected several times – the first time from the actual object in question (hidden from the public) and then the second time from a pane of glass suspended on stage; the object thus appears to float in mid-air on stage. Pepper permitted other objects to be illuminated behind the glass so that the 'ghost' could appear over real objects and which of the two dominated could be controlled as desired by adjusting the relative light levels (Figure 1.16).

This is a form of projection in that there is a light source, an object and an image and in that the end result can be interrupted by means of placing something in the way of the light beam.

This is by no means an effect which is restricted to the Victorian era. The author for example is acquainted with a more recent version which was created by Esquire Jauchem Herbert Senn and Helen Pond for a production of *Der Freischutz* which the author lit for the Opera Company of

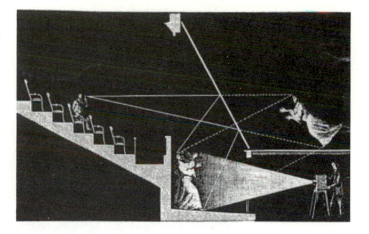

Figure 1.16 Section drawing showing how a real person hidden in the orchestra pit (or offstage) can be illuminated and the reflection appear in a sheet of glass (today plastic) on stage (*Pepper's Ghost*). *Source: Tabs*, 1971, Vol. 29, no. 3, p. 104.

Boston in the USA. In this case a vast plastic sheet was suspended mid-stage and the images reflected from behind a scenic rock. Behind the plastic there was a waterfall effect (plastic tubes containing dry-ice and backlit with water projections) which could be controlled so as not to overcome the 'ghost' image. The plastic sheet rippled as the objects got into place and then became stationary so that the 'ghost' appeared to assemble in place rather than move into place.

Shadow projection

The earliest form of projection was that cast by a shadow and this simple effect was turned into an entertainment initially by the countries of the far east, notably Indonesia where they are still popular. The idea travelled across the world to Europe, 'shadow-shows' were given from the late seventeenth century in Paris and London where they were frequently mounted in the Punch and Judy theatre. More recently people have been introduced to the concept of shadow play through the intricate films of Lotte Reiniger who used tin shadow puppets in the 1930s (Figure 1.17).

Adolph Linnebach developed a form of shadow projection which bears his name. Demonstrating the effect at the Court Theatre Dresden in 1916, he subsequently became the Technical Director of the Munich Opera from 1923 to 1944 and developed many other lighting techniques. The Linnebach shadow technique is the only projection system which (like that of the overhead projector – of which more later) enables the effect to be judged whilst the slide is being prepared *in*

Figure 1.17 Lotte Reiniger's distinctive shadow animation technique, inset a phenakistoscope disc of Chaplin which she designed. *Source*: MOMI, p. 28.

situ. The lack of a focusing lens means that the image is not crisp and some distortion is inevitable if the projector is mounted to the side of and at an acute angle to the screen. Linnebach is most useful on small stages and then only to suggest a background rather than to convey precise information. Some years ago Strand Electric produced a specific Linnebach lantern, the Pattern N623, based on the Pattern 223

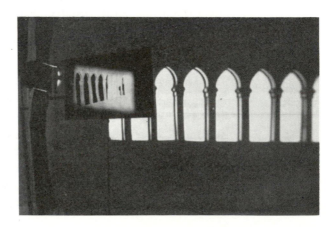

Figure 1.18 A Linnebach lantern showing slide *in situ*

Figure 1.19 Linnebach projection at the Westminster Theatre, London, 1931 for *Tobias and the Angel*. *Source: Tabs*, Golden, p. 39.

1kw fresnel, but this is no longer available (a more sophisticated version which cross faded coloured patterns was available in the later 1930s to early 1940s and was called the Penumbrascope). The beam of modern spotlights is too sophisticated for the effect to work and so the lenses and reflectors have to be removed, fresnels and PC spots are the most suitable (Figure 1.18).

One early use of this technique in the UK was for the production of *Tobias and the Angel* at the Westminster Theatre in 1931 (Figure 1.19). Another production later was praised for its 'realistic impression of the atmosphere of a late November afternoon', the slide being painted by the set designer.

Schwabe, Reiche and GKP

The name Schwabe became well known in Europe in the 1920s and 1930s for a unique system of cyclorama lighting which involved projection and it is in this aspect which we are interested. The Schwabe cloud machine had a single 3 kW light source which was diverted to 2 tiers of condenser lenses, 10 in the lower and 8 in the upper tier. Each lens had its own cloud slide, and two motors, one for each tier, causing the clouds to overlap each other and allowing some control over each tier;

Figure 1.20 Schwabe 3 kW cloud projector. *Source*: Bentham, 1992, p. 316.

the whole construction was also rotated at very slow speeds by a silent motor. The clouds could be from actual photographs or the slides could be substituted for plates made up by the designer or scenic artist. Glyndebourne Opera House was fitted with this device when it opened in 1932 but Basil Dean had used the system at the St Martin's Theatre London in 1923 for his production of *A Midsummer Night's Dream* and again in 1924 at the Theatre Royal Drury Lane. The 1931 London Coliseum production of *A White Horse Inn* also used the Schwabe system. Upon seeing the Coliseum clouds George Bernard Shaw is reported to have banned their use in his productions 'the audience would be so busy staring at the clouds they wouldn't listen to my words'. Indeed there are many such reports of a distracted audience at this time (Figure 1.20).

The cloud machine was developed for Hans Schwabe by a young man called Reiche and most of Schwabe's post-First

World War development was in his hands. Reiche later formed his own company and the Reiche (and Vogel) scene projector became the standard system until the Pani BP4 came along much later (see below). The main rivals in Europe to the Schwabe–Reiche and Vogel systems were Geyling, Kann and Planer from Vienna whose equipment took the trio's initials – GKP. The early GKP machines needed a 100 amp arc to cover the vast cycloramas of the German opera houses, this was subsequently replaced by 60 volt 50 amp tungsten lamps. Interestingly part of the GKP's work in the early 1930s was involved with large scale projections onto the exteriors of buildings, anticipating the Jarre concerts (see Chapter Four) and Pani's BP4 work by many years.

Strand Electric

The famous Strand Electric company has been involved in many early uses of projection but of course they were not the first company in the UK involved in electric stage lighting. The firm of Thomas J. Digby became dominant towards the end of the nineteenth century through his development of the carbon arc which replaced the limelight. Digby had developed an arc slide projector which was called the Pattern 85, and incidentally it was from Digby that Strand developed the 'Pattern' nomenclature for their range of equipment. The carbon arc was to remain the lightsource for projectors for many years.

Strand Electric was founded in 1914 and itself became the dominant company within a comparatively short time. For the Empire Exhibition of 1924 and 1925 at Wembley the company projected a waterfall 25' high and 7' wide (7.5 × 2 m) from a 3 kW lamp, the slide was made from mica and apparently the client thought the joins in the mica were logs falling over the waterfall! By 1933 the company were demonstrating their arc projectors (Pattern 33) which could produce moving flames, clouds and waves. The basic format of the circular revolving disc has changed little since those days excepting that the clockwork spring mechanism has thankfully been replaced! (see moving effects in Chapter Five). In one famous demonstration to George Bernard Shaw the cloud machine did not find favour, GBS was reported as saying it was like the army, 'the same chap came round again and again!'

The original effects were made up of four hand-painted pieces of mica and it was not until after the Second World War that heat resistant glass was used with a combination of photography and art work by the doyen of effects men, Eddie Biddle, who looked after the equipment at Strand and later when the inventory was sold to Theatre Sound and Lighting (TSL).

By 1937 the Pattern 51 Optical Effects Projector was a standard item, powered either by a 1 kW lamp, the range of moving effects was extensive and many will be familiar to the reader as the names of current effects: Fleecy clouds, storm clouds, rain, snow, waterfall, running water, smoke and flames, sand storm, rough sea waves, water ripple, jazz colour and dissolving colour. Later versions were employed during the Second World War to project clouds and waves onto the cyclorama of a Torpedo Attack Teacher in 1942. The centrepiece of this system was a specially designed epidiascope which projected images of enemy ships which were in fact scale models picked up by the epidiascope. The instructor had pushbutton control of motorized dimmers which delivered a variety of weather conditions, seasons and times of day. This system anticipated the flight simulators of today's aircraft which also use reflections and projections (of computer images).

Strand were ingenious when it came to problem solving, a typical example is one from 1938 when the company were called upon to project clouds and an aeroplane in the GPO pavilion of the Glasgow Exhibition. This extract from the company's publication 'TABS' explains the problem and the solution, and allows us a very clear picture of the technology available at that time:

> The entire ceiling in the picture had to be covered with moving clouds and a projected aeroplane made to travel right round the walls, crossing the oceans and traversing enormous cut-outs of the Dominions and the British Isles. . . . the job was complicated by the fact that the 'plane had not to appear on the columns which were spaced to an elliptical plan. Nor could the projector be placed centrally as this position was occupied by a lattice mast which went right up through the roof. The solution was to use two synchronised projectors with U.V. light sources. Thus only where a fluorescent track was painted would the plane appear. If this principle were accepted the clouds overhead might just as well be fluorescent also. The first hurdle that had to be overcome was the projection system because the only sources available for U.V. were 125 watt black lamps! For the clouds, Stelmar reflectors were used because these would collect from the front of the lamp. For the plane a narrow angle condenser system with a stencil in the gate was employed. Only twelve of the sixteen projectors would have moving cloud discs (photographed on glass for the first time) and the rest were to have static cloud photographs on glass slides 7" × "in size". (Figure 1.21).

Figure 1.21 One of the 16 special ultra violet projectors used by Strand in the GPO pavilion Glasgow in 1938. *Source: Tabs*, Golden, p. 92.

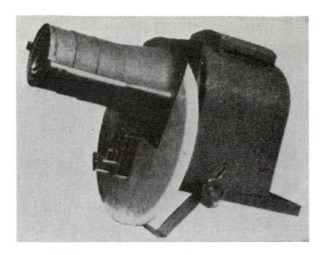

This extract is also one of the first references to the gobo, called 'stencil' in the text. It was not until the early 1970s that Strand, by then Rank Strand, decided not to make large scale scenic projectors but to import others and the company thus took on the Pani agency for a few years. Prior to this Strand manufactured the 4 kW Pattern 152 which later became the Pattern 752. There was also a Pattern 452 and the 2 kW Pattern 252 which most recently has become the 2.5 kw Cadenza FX, the machine used for the projection of the moving discs reminiscent of those used in the 1930s.

Ludwig Pani

In recent years the name Pani has become synonymous with large scale scenic projection, perhaps taking over from Reiche and Vogel who had held the position earlier in the century.

The company was founded in 1930 by Ludwig Pani, and in the early days it was the scientific rather than the theatrical which formed the basis of the business. It was not until after the Second World War and with the reopening of the Vienna State Opera that Pani went into scenic projection in a big way with the then new lightsources and wide angle lenses being the catalyst. The company's core product is the BP4 which was introduced at the Bayreuth Festival in 1973 and quickly became popular especially throughout European opera houses. The BP4's main advance was through the use of the 4 kW halogen metal vapour lamp, smaller versions of which were by then enabling follow spots to compete and surpass the

Figure 1.22 One of Pani's range of scene projectors, the BP5 5 kW halogen fitted here with f 13.5 cm lens. Photo courtesy of Ludwig Pani.

performance of carbon arcs. The HMI and CSI lamps cannot be dimmed by varying the voltage and thus mechanical means had to be developed. Pani produced servo operated glass sliders which were gradually shaded to black and which were said to duplicate the dimmer curve of the thyristor (Figure 1.22). It is interesting that in 1899 George Applebee, who was the electrician at the Gaiety Theatre in London, patented a dimmer which employed sandblasted mica.

Summary

Most of the basic laws of optics were developed thousands of years ago, and the work described in this book demonstrates that the last four hundred years have been more concerned with the application of those laws rather than with totally new discoveries. It is possible to postulate that the Ancient Greeks could comprehend a modern slide projector because its optics are based on the laws they themselves used. Only new developments which the ancients could not have foreseen (such as lasers – although some people do argue that these too were predicted) have moved projection into truly new realms of possibilities. Much of the early developments were echoed across continents as projectionists came to the same realization about the same particular problem at the same time, but were prevented from learning of this by the primitive, and time-consuming, communications of the period. Only later was it possible to learn from each other more quickly, with consequent acceleration in the sophistication of the equipment. Another interesting factor is the influence which medical work had on the development of projection as physicians used the new medium to explain their latest work, often using live models, similarly the magic lantern brought photographs of news items to its audiences. The Victorian period in particular was a rich period and saw the components of projection utilized in toys and other almost domestic pieces of equipment for everyone's entertainment; this alongside formal presentations of projection either alone, or in conjunction with, other aspects of performance. By the turn of this century the fascination with the moving image had reached such a point that the development of still projection had slowed considerably, especially in the UK and this contrasted with the work in Germany and Austria where large-scale scene cloud and projectors were needed for the big stages of the opera houses, even today the best and largest scene projectors are made in Austria. If there is a trend in this century it is the expanding use of these machines (and their smaller counterparts) out of the large opera houses and state theatres and into smaller commercial and regional theatres.

2
The equipment

Slide projectors

For most types of projection used in the theatre, the delivery
system will be a slide projector and all types observe the same
physical characteristics irrespective of size. The basic components are listed here and see also Figure 2.1.

- the housing for the lightsource;
- the condenser lens system;
- the locating mechanism for the slide (known as the 'gate');
- the objective lens system.

The condenser lens system is there to evenly distribute the
light over the surface of the slide and the condenser is usually
contained within the same housing as the lightsource itself.
The housing will also contain heat absorbing glasses and filters
to remove any part of the optical spectrum which is created
by the lightsource and which might damage the slide. Most
slide projectors, even comparatively small ones, contain a fan
cooling system to lower the temperature on the slide and
protect it from the heat generated by the lightsource.

Large slide projectors tend to have lightsources which
cannot be dimmed by conventional means and thus they are
provided with mechanical dimming devices. These have
distinct advantages as we shall see later.

Most slide projectors provide some system of changing the
slides from a remote location, and generally the smaller the
projector, the more slides it can hold at any one time. Thus
smaller projectors tend to be chosen not for their light output,
but for their ability to deliver a large number of images in
quick succession and with a high degree of reliability.

Figure 2.1 4 kW slide projector
with side removed showing
lightsource, condenser lens system
and heat absorbing glasses. *Source*:
Bentham, 1992, p. 68.

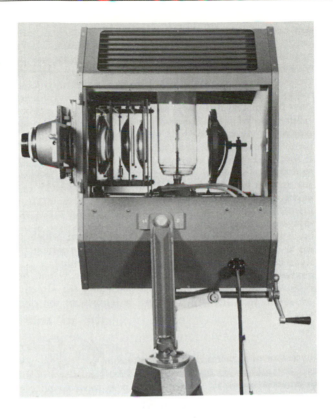

Figure 2.2 35 mm projector from
Kodak, the Ektapro, which replaces
the ubitquitous Carousel. Photo
courtesy of Kodak.

Conversely the larger projectors tend to be chosen specifically
because their power and format enables them to cover large
areas with ease, and many of these too are now available with
slide magazines. Professional (as opposed to domestic) slide
projectors can usually be connected to controls which will
select any slide not necessarily those in chronological order –
known as random access (Figure 2.2).

Objective lenses are selected for a specific project and usually smaller slide projectors provide a greater range of objective lenses than the larger ones; thus small projectors are more flexible in terms of location, throw and picture size. Small projectors can also be provided with zoom lenses which are very useful in touring. Large scene projectors used for theatre work tend only to use zoom lenses for effects.

Overhead projection

It would be a mistake to relegate overhead projectors (OHPs) to the lecture room because the OHP of today can be connected to very sophisticated computer generated slide devices. Alternatively the production of material for this format is within the reach of most people with a personal computer and a local copyshop. The OHP follows the format of a slide projector in that there is still a lightsource, condenser, gate and objective lens but they are arranged slightly differently. The main difference is that the objective lens has limited focus and so the distance from the screen that the OHP can be placed falls within a small range if the image is to be in focus.

Although there are OHP machines with high powered light-sources for large screens they are very rare and most readers will be more familiar with the small device which projects a beam about 2 m² at the most. One drawback for theatre work is that the device has to maintain a specific relationship with the screen and thus can often require hiding behind some

Figure 2.3 Elite optics LCD tablet projecting computer created images via the OHP. Photo courtesy of Elite.

scenic device; although it is also possible to back project with OHP as well. The screen must be at right angles to the beam centre line if keystone distortion is to be avoided. One additional asset of the OHP is its ability to project a continuous role of acetate and this can be used on stage to simulate some degree of constant movement.

OHP's can now be provided with a liquid crystal display (LCD) tablet which connects to a PC and which will enable the lecturer to write his or her work at home and then deliver an almost animated sequence of complex text and graphics at the touch of a button. The possibilities for theatre work, especially in small venues, are obvious (Figure 2.3).

Profile spotlights (lekos in USA)

Profile spotlights (lekos in the USA) are part of the normal day-to-day inventory of stage lighting but they offer the additional facility of projecting shapes because the profile spot is essentially a simple projector. The unit consists of an ellipsoidal reflector (hence the name 'ellipsoidal' in the USA), a gate and an objective lens system which in more recent generations is likely to have two components so as to offer a variety of beam angles (colloquially known as 'zoom'). The lekos produced in the USA have fixed beam angles and thus specific instruments must be chosen for specific beam diameters. The gate will accept metal or glass slides (known as gobos, or 'patterns' in the USA) and it is these which are relevant to this book as later sections will describe. The projected pattern is either held in clear focus if a precise image is required or it is held out of focus if overlapping patterns are needed to add some texture to some scenic surface. The burning position of the lightsource relative to the reflector can be adjusted so that the light beam is either 'flat' or even, or 'peak' in which case the light is concentrated in the centre of the light beam. This latter adjustment can make a considerable difference to the light output and quality of the result and readers will be wise to select spotlights which offer a good degree of control (preferably via hand-held knob rather than screwdriver) over this facility. This technique combined with split, or broken, pieces of colour can achieve a high degree of sophistication, and yet is within the reach of any theatre possessing this everyday piece of equipment. Future profiles are likely to have some variation on the LCD device now common on OHPs so that a disk from the manufacturer will provide a wide variety

Figure 2.4 A selection of gobos from Great American Market made for the Intellabeam and Trackspot moving lights, these precision gobos are only 1″ in diameter. Photo courtesy of Great American Market.

Figure 2.4 A selection of gobos from Great American Market made for the Intellabeam and Trackspot moving lights, these precision gobos are only 1″ in diameter. Photo courtesy of Great American Market.

of gobos at the touch of a PC key and perhaps to enable the designer to draw their own via lightpen or graphics tablet (Figure 2.4).

Lamp history

Lamps tend to be named after the method by which they produce light. Hence, since the filament in incandescent lamps is made from the metal tungsten, it is this latter word which is used to describe this particular family. Tungsten has the highest melting point of any element, lamps containing tungsten filaments were developed in the nineteenth century. Until the 1960s tungsten lamps were large, in order to keep the glass sufficiently far away from the hot filament to prevent it melting. The introduction of both a halogen element into the gas within the glass envelope, and the simultaneous introduction of borosilicate into the glass made it possible for the lamps to be smaller and also prevented the filament from evaporating and depositing itself on the inside of the glass wall, thus reducing the light output. (This deposit increases with an increase in the temperature of the filament.)

The current version, which appeared in the early 1970s has developed further and provides considerably extended life by comparison with the lamps which were available only twenty or thirty years earlier. This is possible through the 'recycling' effect of the halogen which, together with the increase in the pressure of the gas and the reduced size of the lamp causes many of the metal particles to move back to the filament from the glass wall of the lamp.

In addition to the benefits of maintained light output and longer life, the tungsten halogen lamps also became much smaller and this in turn enabled luminaire and projector designers to make their lamp housings much smaller. One consequence was that the increased heat output from the lamp was contained within a smaller reflector housing and the designers had to work considerably harder to produce a luminaire which was not only smaller and brighter but also cooler. Of course most projectors are also fitted with fans which cool the gate where the slide is presented.

The design of the filament influences the quality of the projected picture as we have mentioned above. Early tungsten lamp filaments were 'monoplane' in other words the strands of the filament were all in one line. 'Bi-plane' lamps bunch the strands of the filament much closer together in two rows and this means that the filament can be more compact, thus producing a clearer image. The compact filament also enables the light to be gathered more efficiently by the reflector and lens system. Some lighting designers prefer profile spotlights to be fitted with bi-plane filament lamps to project cleaner gobos.

Projector lamps have advanced ahead of those used in theatre luminaires, for example projector lamps are often fitted with reflectors inside the glass envelope so that not only is the orientation accurate each time, but also every new lamp provides a clean, and not burned, reflector. Theatre spotlight lamps came to this concept much later.

Lamp types

Technically (as opposed to creatively) the most effective projections are those which are clear, undistorted and of sufficient, evenly distributed intensity to be accurately balanced with the rest of the overall design.

These requirements are all delivered by a specific approach, each of which in turn is influenced by the laws of optics and the characteristics of the main components. Clarity, for example, is much influenced by the size of the lightsource, not so much of the envelope of the lamp (although of course this has an effect on the size of the reflector, lamp house and so on) but the size and shape of the actual point of light – the more compact the better.

The type of lightsource selected has a major influence over the resulting picture and, in turn, the key performance factors of sources are their intensity, stability, dimension and ability to dim (although a source would not be ruled out if in all other respects except ability to dim it met specifications, since mechanical dimming is now very sophisticated as we shall see later).

Of the higher powered light sources, HMI and Xenon are the most common, both rely on the arc principle but in the HMI there are two arcs and thus a point source of light is not possible. When this is required xenon, which has a single arc, is the more preferable. The width of the arc in a xenon is typically 4 mm (in 5 kW and 7 kW lamps) and as such can be described as a 'point' source by contrast with the 30 mm wide arc of the HMI which is known as a 'spread' source.

Additionally because HMI lamps work on alternating current (AC) they have a tendency to flicker which, whilst often unnoticeable to the naked eye, is nevertheless picked up by film and television cameras. Xenon, by contrast, works on direct current (DC) and therefore does not flicker. HMI, however, is the more efficient of the two in converting electrical energy into light – typically twice the light output per watt by comparison to xenon.

Despite this, most recent slide projectors are xenon, a selection which has been made easier because of the development of more efficient current rectification. Until recent years it was only cinema projection which could accommodate the large rectifiers required. Now xenon machines are portable and do not require anything like the large currents which they once did.

Each type of lamp requires a different reflector and lens system. Large xenon lamps work within an ellipsoidal reflector and bounce their light off a dichroic cold mirror onto the condenser lens through additional cold filters to remove the infrared portion of the spectrum, this enables a wider range of film to be used than is possible with the HMI lamp. Because of the design of the light source and reflectors, the main portion of the xenon beam can be better directed towards the centre of the condenser lens which reduces, and in some cases even avoids, the spherical and chromatic aberrations caused by the edges of lenses (Figure 2.5).

By contrast, the HMI source requires a small spherical mirror, working to a fresnel lens (not to be confused with the fresnel lenses in theatre spotlights which are textured on the reverse to soften the edge of the light beam). Because the

Figure 2.5 An ellipsoidal mirror collects the light from the xenon lamp and directs it to the condenser lens via infrared filters which consist of a dichroic cold mirror and a cold filter. The shaded box represents the slide.

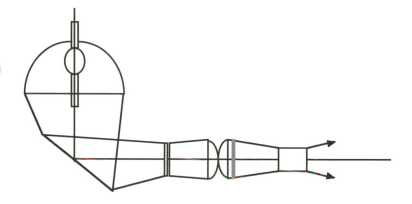

mirror does not surround the lamp, much light is lost and what light is reflected tends to overheat the lamp which shortens its life. Thus xenon light sources are the more efficient (Figure 2.6).

Another facet of the xenon is its ability to permit the current to be adjusted so that the intensity of one projector can be matched to that of another, a major requirement of multi-screen presentations. HMI lamps do not permit this. Additionally xenons deliver virtually the same light through-out their life (for the same current) whereas HMIs suffer much loss of light as they are used, partially because of the overheat-ing caused by the spherical mirror.

The spread of the arc influences the quality of the resulting picture when used in conjunction with different objective lenses. For example the 'spread' light source of the HMI works best with short focal length lenses although these lenses have a tendency to concentrate the light towards the centre of the screen and thus it is wiser to adjust the lamp with respect to the reflector so that a more even, flat distribution is obtained, even if this results in less light output overall. 'Point' sources

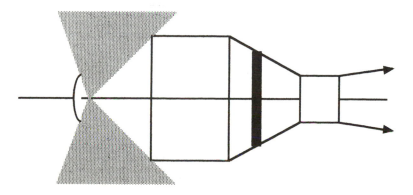

Figure 2.6 In the HMI light system much of the light is lost and additionally some light is reflected back through the lamp by the small spherical mirror which reduces the lamp's efficiency and life.

Figure 2.7 Light sources and lenses.

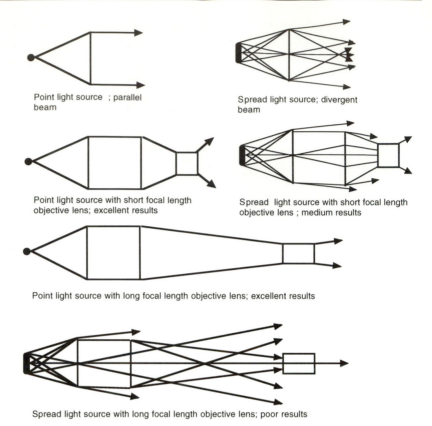

Point light source ; parallel beam

Spread light source; divergent beam

Point light source with short focal length objective lens; excellent results

Spread light source with short focal length objective lens ; medium results

Point light source with long focal length objective lens; excellent results

Spread light source with long focal length objective lens; poor results

such as those found in xenon lamps work well with lenses of all focal lengths (Figure 2.7).

Colour temperature

Incandescent lamps must obviously be designed not to burn at such a temperature that the filament itself melts. Conversely the lamp needs to burn at a high temperature in order both to produce a high light output and to produce as white a light as possible. If we assume that the colour of sunlight is white, as a point of reference, then a filament would have to burn at the temperature of the sun (6000 degrees Kelvin) in order to produce the same colour of light. As we saw above, tungsten is the metal used in lamp filaments because it has the highest melting point (at about 3410 degrees Centigrade ± 20 degrees). The colour which the filament produces matches the colour spectrum as its temperature increases, hence at lower temperatures the filament produces a large quantity of infrared (which is filtered out in projectors as we saw above) whilst at temperatures higher than those of the sun the filament would produce ultra-violet. The colour temperature of the projector lightsource must be noted by the person responsible for photographing the artwork and producing the slides as it could influence the choice of film.

Another factor for lamp designers is the life of the lamp, generally the light output can be increased if the life is decreased and thus we can see that for projection, which would usually require a high light output, we could accept a lower life than would be acceptable in, say, a theatre spotlight.

The section on dimming later in this chapter contains a useful table on how the colour temperature of tungsten halogen lamps alters as the dimmer is adjusted.

Types of lenses used in projection

The basic lens is plano-convex, that is, it is curved on one side and flat on the other. These lenses can introduce distortion and this can be reduced by curving the lens on both sides in which case it is known as a 'bi-convex' lens. The same effect can be created by placing two plano-convex lenses with the curved or convex sides very close together, the familiar format of the condenser lens. The distance between the lenses influences the overall focal length of the system, if they touch then the focal length will be considerably shorter than if they are moved apart (at which point they will also have to be moved further away from the slide) and this is the principle behind the zoom lens. The longer the focal length becomes, then the higher the 'f' value and the less light the system will pass (Figure 2.8).

The process of passing light through glass lenses involves refraction, the bending of the light caused by it travelling at different speeds in different media. Not all parts of the colour spectrum are refracted the same way and so in a simple lens some colours become more visible through being separated – known as 'chromatic aberration'. Another problem with simple lenses at wide angles is the 'pin cushion' effect, a somewhat curious description of the curvature of the projected image away from the sides of the screen. Both these problems are overcome by making lenses of not one but several glass components, many ground so that they will refract in different ways, the net result being no coloration or distortion of the picture. Special coatings also reduce refraction problems and today the coating is vacuum applied in order to reduce the

Figure 2.8 Lenses: left, plano convex; centre, condenser lens; right, simplified version of the Fresnel lens.

Figure 2.9 A simplified diagram of a compound lens which is comprised of several different lenses joined without an air gap, shown here for clarity.

light loss which could be the result of air getting into the gap between one part of the lens and the next (Figure 2.9).

Fresnel developed his lens in the nineteenth century for use in lighthouses where it remains. The main benefit of the fresnel lens is that it reduces the amount of glass required by the plano-convex lens which its shape echoes. This is possible because the lens is a series of concentric prisms which surround a small convex lens. The soft edge associated with fresnel theatre luminaires is produced by the textured surface on the rear of the lens, a feature also of the PC lenses also produced in recent years for theatre luminaires. One manufacturer engraved prisms and since the lens shape is (plano) convex the description prism-convex led to the initials PC. This is not to be confused with the initials PC which also stand for plano-convex which is also the shape of the prism-convex lens!

How lenses work

For both projection and photography, the more light the lens allows through the better the final result will be. The amount of light passed is related to the f value of the lens which is the ratio of the focal length of the lens to its diameter (technically to the aperture of the shutter opening in a camera, or the width of the slide in projection – but effectively these will be the same as the effective lens diameter). The smaller the f number then the more light is passed and thus for projection low f lenses are the most desirable. Low f lenses are sometimes called 'faster' lenses.

Thus a lens which was 100 mm diameter and which had a focal length of 300 mm would have an f value of 3, but if the diameter of the lens was only 50 mm then the f value would increase to 6 and thus the lens would not pass as much light. This illustrates why lenses have to be large in order to have a good f value. Long focal length lenses are used to project over long distances and short focal length lenses over short distances. 35 mm slide projectors normally use lenses with focal lengths between 60 and 120 mm with special requirements involving lenses outside this range at focal lengths typically of 35 mm, 250 mm and 300 mm. The larger 180 mm and 240 mm slide projectors also have a wide range of lenses typically in the case of Pani extending up to 600 mm. When the projector is very close to the screen a wide angle (short focal length) lens is needed, wide angle lenses have a more prominent curve than long throw lenses. If our desired f value was 3 then a lens with focal length of 60 mm would have to be 20 mm in diameter to achieve this. At a focal length of 120 mm, the lens would have to be 40 mm in diameter in order still to deliver an *f* value of 3. Thus it becomes clear that lenses with long focal lengths (that is for long throws) will tend to have higher '*f*' values than short focal length lenses because otherwise they would be too big and furthermore since the light beam in slide projectors tends to be concentrated towards the centre of the lens, there is little point in having larger diameter lenses anyway. Note the intensity of the picture is not just governed by the *f* value of the lens – light intensity, slide size and quality, picture size, throw, screen type and ambient light all have their influence.

Diameter of lenses

Today most people wanting to project on stage will use some proprietary equipment, probably hired for the specific project in hand. Sadly, perhaps, people are less inclined to experiment today than in the past. For those who do it is useful to remember that lenses are designed to relate to the size of image which they are 'seeing' in the sense that the length of the diagonal of the slide approximates the diameter of the lens. Should a lens be used, the diameter of which is significantly less than the diagonal of the slide, then the picture will be dark at the corners. Conversely a lens larger than the slide could cause some scattering of light and cause the contrast to be adversely affected.

Prism lenses

Perhaps originally associated with the kind of effects projection seen in discos, prism lenses are now available for quite large scenic projectors. The lens is actually an attachment to the normal lens and it causes a specified number of overlapping images to be formed (Figure 2.10).

Figure 2.10 A prism lens fitted to a Pani projector. Photo courtesy of Ludwig Pani.

Depth of field

'Depth of field' is the term used to describe the distance over which the lens will produce a sharply focused image, lenses usually provide greater depth of field behind the ideal point

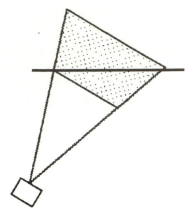

desired for the picture (known as the critical focus) than in front. Wide angle (short focal length) lenses have greater depth of field than narrow angle (long focal length) lenses. This is fortunate for theatre work since wide angle lenses are more likely to be used, especially from acute angles in the wings and thus the greater range of focus of these lenses is useful considering the distances involved. The depth of field for a given optical system can be calculated but this is unlikely to be required for theatre work; nevertheless most books on optics will provide the necessary equations should the reader like to take this aspect further (Figure 2.11).

Figure 2.11 The shaded area represents the depth of field, the distances within which the image will be in focus, a good depth of focus ensures that the image will be sharp when at an angle to the screen.

Dimming

The tungsten and tungsten-halogen lamps associated with stage lighting can be dimmed by varying the current through the use of the thyristor dimmer or through older methods involving wire-wound resistance types and transformers, but lamps which require to 'strike' such as HMI and xenon cannot be dimmed by these methods and require a mechanical system of controlling the level of light emerging from the projector. Thus manufacturers of large projectors which require HMI or xenon lightsources usually supply mechanical faders with their machines, either in-built or available as an optional extra (Figure 2.12).

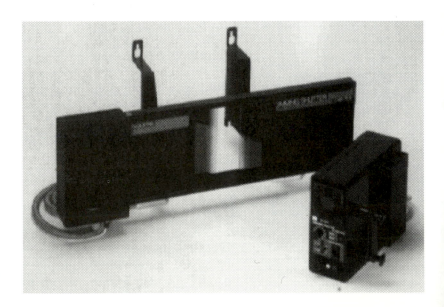

Figure 2.12 Pani processor controlled grey scale dimming shutter. Photo courtesy of Ludwig Pani.

Whilst it might appear cumbersome and crude in this day
and age to fade light by mechanical means, in fact there are
distinct advantages in doing so. In comparison with dimming
through voltage and current control – the technique used in
stage lighting – mechanical dimming leaves the lamp burning
at full intensity and thus the colour temperature does not
change during the fade. This is a factor of dimming tungsten
and tungsten-halogen lamps. For example if such lamps
are dimmed through a linear thyristor dimmer then the
colour temperature varies with the fader setting as shown in
Table 2.1.

Table 2.1. Colour temperature in relation to fader setting.

Fader setting on lighting desk	Light output as percentage	Colour temperature in degrees Kelvin (K)
10	100	3200
9	81	3120
8	64	3040
7	49	2960
6	36	2860
5	25	2750
4	16	2600
3	9	2400
2	4	2200
1	1	
0	0	

Perhaps the most sophisticated mechanical fader is that
produced by Pani in which grey glass is moved gradually by
stepper motors which permit the action of fading to be
controlled from lighting desks with considerable accuracy. The
fader can operate at speeds from 0.35 to 30 seconds and it can
be fitted with a fan when it is installed in the higher powered
projectors.

35 mm slide projectors such as the Kodak Carousel or the
newer Ektapro cannot simply be plugged into a dimmer even
though the tungsten halogen lamp can be dimmed this way.
This is because the main power supply cable also serves the
slide changing mechanism and the cooling fan both of which
must be maintained at all times. Thus these machines have
accessory sockets into which special dissolve units can be
plugged. The dissolve units offer a selection of fade rates in
seconds and so do not permit the balancing of levels in the
sense that stage lighting control does (Figure 2.13).

Figure 2.13 A Kodak dissolve unit
for 35 mm slide projectors. Photo
courtesy of Kodak.

Front projection screens

Almost any surface will reflect light to some extent, even a
wall painted matt black and a length of black velour thus, in
theory, almost any surface will be acceptable as a surface for
projection. Some readers might be acquainted with the Las
Vegas and London Palladium revues, frequently starring the
pianist-showman Liberace, in which a battery of projectors
would be focused onto a wall of water! However, only
purpose-made screens are likely to be the most efficient in
terms of reflecting the light back to the viewer at the correct
angle.

A screen is required to have little adverse effect on the
brightness of the picture and on the distribution of light within
it, and also on the colour of the picture. The gain of a screen
can be measured and the figure obtained indicates the gain of
the screen in question in comparison with a reflecting surface
which diffuses light equally in all directions.

A matt screen will appear uniformly bright from whichever
angle it is viewed and so this is the best type to choose if the
audience are seated at a wide angle in relation to the screen.
Semi-matt screens are available with a gloss finish and these
have a higher reflectance in the centre (and thus a higher
gain) but at the expense of those situated at acute angles.
Semi-matt screens should therefore not be used in wide
angled auditoria. A typical value of gain for a matt screen is
0.85 and 1.5 for semi-matt. Higher values are possible from
screens with metallized surfaces where metallic paint has been
used to increase the reflection even further. These screens are
very directional but have gains in the order of 2.5 and it is

this type of screen which is used in commercial cinemas where the viewing angle is very defined. The tables in the Appendix demonstrate the difference between the Harkness matt white, which maintains the light level relatively well at increasing degrees off axis, but compare this with the Harkness 'Perlux' high gain screen which has almost the same pattern off-axis as the matt, but more than twice the gain on axis. Harkness also manufacture a range of special high-gain screens in which the gain can be in excess of 4 but at the expense of the viewing angle, Harkness advise that in this case it is restricted to within 25 degrees off axis. Some projection experts fit reflective aluminium foil behind their screens to increase the reflectance.

Rear projection screens

Front projection involves a number of potential problems. Unless the projector can be located in a suitable room at the rear of the stalls or the circle, then some distortion is almost bound to occur, this problem could be acute if the projector is located at the side of the stage. Additionally the front projection beam prevents scenery or actors being located within it and thus creates a 'no-man's-land'. Rear projection, however, brings some considerable benefits (Figure 2.14).

First, rear projection clears the performing area of any light beam and so scenery and actors can be located wherever desired. Second, it is easier to place the projectors on the centre-line of the screen and so avoid distortion. Finally the projector is well backstage, where it is not only easily accessi-

Figure 2.14 Front projection creates a cone of light which actors and scenery cannot usually enter, rear projection removes this obstacle and provides better access to the projector.

ble in case of problems, but also in a location where much of the sound of the cooling system is absorbed.

When planning rear projection the critical factor is the range of lenses which the chosen projector can accommodate because most stages are not deep enough to house both a good sized set and a long throw for the projector, and so wide angle, short throw lenses are vital. Occasionally the depth available is insufficient and so the beam has to be diverted via a mirror system. The arrangement can be calculated by making a simple template out of paper and then placing the template on the plan of the stage and folding the template over where the beam crosses the rear wall of the stage. It is at this point that the rear mirror will be placed and, assuming that the template is to scale, then the measurement along the fold will provide the width of the mirror (Figure 2.15).

Domestic mirrors are silvered on the rear and if used for reflection as above then a small amount of light will be reflected from the glass and will create a slight ghost of the main image. This is unlikely to be a serious problem since in most situations even the nearest spectator is a considerable distance away, but if the projection distances are

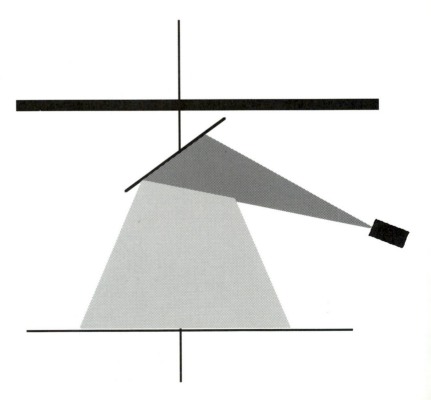

Figure 2.15 Bouncing the projection beam off a mirror enables the throw to be contained within a restricted stage depth.

themselves large, and the image unclear to begin with, then the error can be magnified and visible, especially if the event is being televised when cameras may work close to the action. Surface silvered glass mirrors are available – they will overcome this problem but they are expensive. In many cases the shrink mirror fabric is adequate and this has the added bonus of being lightweight and less prone to damage than the glass types, although of course the surface can easily be pierced.

Any material which is translucent can be used for rear projection but there are two factors to take into account when choosing a screen. The first is the transmission factor of the material and the probability is that only a purpose-made back-projection material will have the highest trans-mission, these materials will also have a matt surface to avoid reflections from the front. The other factor is the viewing angle, and the angle within which the intensity of the picture remains acceptable – the darker the screen material then the narrower this angle. For example black back-projection material, universally accepted as a most magical material because unlit it does not look like a screen, has only a 30 degree viewing angle and thus people outside this cone, to the sides of the auditorium perhaps, will see an image much reduced in intensity. By contrast the 'Twin White' material can be viewed anywhere within the 180 degrees of the face and can accept both front and rear projection (Figure 2.16).

Some materials, especially those used for back projection have a tendency to create a hot spot where the viewer 'sees' the projector itself, more recent proprietary materials have in-built diffusers to minimize or eliminate this problem.

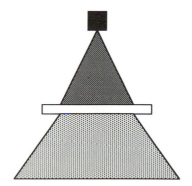 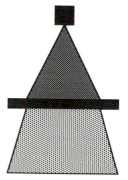

Figure 2.16 The white rear projection screen on the left permits more of the audience, seated at a wider angle to the screen, to enjoy a high intensity picture, whereas, the black (or grey) rear projection screen on the right restricts the angle within which the intensity is high. The audience seated outside this narrower cone will not see as bright a picture as those within. (The picture size is the same in both cases.)

Summary

The only two items which are common to all forms of projection are the source of light, and the object which is being projected. Most projection systems require lenses to focus the picture correctly, but not all do so, for example, shadow projection does not use any lenses at all, and is thus the simplest and cheapest kind of projection with which to work. Lens systems come in two forms, the condenser, the purpose of which is to evenly distribute the light, and the objective, the purpose of which is to focus the picture sharply. It is the objective lens which can be changed to provide different picture sizes at a given throw.

There are three types of projector which require lenses, slide projectors, overhead projectors and theatre profile (ellipsoidal) spotlights. Each serves a slightly different purpose, the slide projector being the most common for projecting large or small scenic elements, the overhead for communications and small pictures, and the theatre spotlights for projecting (often sophisticated) shapes. The first two can be fitted with slide changing equipment, operable from a remote location (frequently via the stage lighting control or similar computer). Conventional theatre spotlights cannot yet be fitted with mechanisms to change their 'slides' (gobos or 'patterns' in the USA). All three projectors would be associated with the projection of still pictures, although they can be fitted with accessories to cause the picture to appear to move to some degree (see below).

The lamps used in projectors determine the clarity of the image, by virtue of whether their filaments form a 'point' source of light, or a 'spread' source (which is inferior), and whether the operating current is steady, or flickers. Additionally the temperature at which the lamp burns has a direct relationship with the colour of the light produced, which, in turn, influences the choice of film chosen for the slides. Not all lamps can dim by varying the current through the use of the thyristor dimmer, and so they are fitted with mechanical dimming shutters which can be operated from the lighting control and follow the dimmer scale.

Light can be projected onto any surface, but a proprietary material is more likely to result in a clearer and brighter picture. Front projection screens are made with a variety of surfaces to suit the angle from which they are viewed and, in a similar way, rear projection screens have varying width of viewing angles – the darker the colour of the screen material, the narrower the viewing angle.

3
Planning

Photometry

When measuring the amount of light involved in stage, television or film lighting, and in projection, there a number of ways in which the measurements can be made. The light output of the lightsource itself can be measured and this would be independent of any optical system which surrounds it. Next, the output of the whole luminaire or projector can be measured, and this would depend on a number of factors – how the fitting was focused, its efficiency, and the nature of the slide involved (if slides are involved at all – manufacturer's measurements cannot take account of slides for obvious reasons). Finally, the light level at the screen or stage could be measured, and this would depend not only on the projector,

Table 3.1

Term	Definition	Units	Application
Luminous intensity (sometimes called 'candlepower')	The illuminating power of a light source in a given direction	Candela (cd)	The output of spotlights (and Pani projectors) is frequently expressed in candela this would be known as the 'beam candlepower' or 'BCP'
Light flux (sometimes called 'luminous flux')	The amount of light falling on unit area	Lumen (lm)	The output of some projectors is frequently expressed in lumens
Illuminance (sometimes called 'illumination intensity' or 'illumination')	The luminous flux per unit area falling on a surface	Lux (lx) = 1 lumen per square metre Foot candle = 1 lumen per square foot; (1 foot candle = 10.76 lux)	These are the usual terms for expressing and measuring the 'target' amount of light falling onto a projection screen
Luminance	The amount of measured brightness, i.e. the amount of light reflected from a surface	Candela per square metre Apostilb (asb) = lumen per square metre. Candela per square foot Foot-lambert	These are the usual terms relating to the 'gain' of a screen, i.e. its ability to reflect light back towards the audience

but also on the distance the projector was from the screen, and thus the resulting size of the image. In theatre work we are most concerned with the output of the projection system, and with the level of the resultant picture; theatre lighting designers and technicians do not normally work with these measurements, which are, however, a key component of the work of film and television people. Nevertheless, the measure of light falling on a surface, under a given circumstance, is a useful piece of information to have for reference (see Table 3.1).

Equations

There is a mathematical relationship between the various elements of projection and thus calculations are possible for various aspects. It should be noted that projection is quite possible without having to make any such efforts, and a lack of skill with a calculator should not deter anyone from attempting projection. If no calculations are undertaken, then some physical tests, on either models or full scale environments, should be seriously considered before the production gets underway. In either case the end result will be more certain for having done some preparatory work.

In the following equation the terms are;

F = desired focal length;
O = width of slide;
D = distance from projector to screen (throw);
B = picture size.

thus, by substitution, any of the above can be found by the following:

$$F = \frac{O \times D}{B + O} ,$$

$$B = O \left(\frac{D}{F} - 1 \right) ,$$

$$D = F \left(\frac{B}{O} + 1 \right) .$$

For example, if the size of a slide aperture is 35 mm and a 4 m wide picture is required from a 4 m throw then the desired focal length of lens is given by:

$$F = \frac{O \times D}{B + O} = \frac{35 \times 4000}{4000 + 35} = 34.69 \text{ (i.e. a 35 mm lens).}$$

The above data might be associated with projection from a Kodak Carousel (or Ektapro machine) using 35 mm transparencies. For larger scenic projections, perhaps those associated with the equipment of Pani, the aperture size of the 180 mm slide is 168.6 mm × 168.6 mm for glass manually changed slides and 155 mm × 155 mm for metal film holders used in automatic slidechangers. Thus the equation for glass slides at 8 m throw, 10 m picture size is:

$$F = \frac{O \times D}{B + O} = \frac{170 \times 8000}{10\,000 + 170} = 133.7 \text{ mm (nearest lens actually is}$$
$$13.5 \text{ cm).}$$

In reality both Messers Kodak and Pani (and other manufacturers), and many of their agents world wide, publish projection tables, (copies of which are published in the Appendix) obviating the need for these calculations most of the time. Nevertheless, the equations are a useful aid, especially for unusual situations.

There is a relationship between the output (luminous intensity) of a projector, and the amount of light (illumination intensity) which will fall onto a surface, and this can be given by the following equation;

$$\frac{\text{illumination intensity}}{\text{(in lux or lumens per sq m)}} = \frac{\text{output of projector in candelas / lumens}}{\text{(distance between projector and screen in metres)}^2}$$

NOTE: only accurate when screen perpendicular to axis of projector.

Obviously, the slide itself has a major influence, but this equation can be used to check the potential brightness of the picture, and to compare the efficiency of both projectors and lenses. It is useful to do the above calculation for a known spotlight for the throw involved, and compare that figure with the figure for the projector. For example, a 1 kW theatre spotlight with a beam angle of 20 degrees will produce 426 lux at a throw of 10 m. Compare this, for example, with the output of a Kodak Carousel with 250 watt lamp and 60 mm f 2.8 lens, which would give a 2 m wide picture at a throw of 3.4 m with an illumination intensity of 375 lux. This is the kind of relationship which the Carousel might have in a lecture situation and, perhaps, as much as one would ask of this piece of equipment, unless used in relative darkness and with a relatively simple slide, in which case it could be pushed to a larger (and thus dimmer) picture.

It is worth noting that the high-powered slide projectors which utilize xenon arc lightsources, produce typically 4000–6000 lumens, which means that if the target illumination intensity is 400 lux, then the total area of projection which these systems could serve at this illuminance is given by:

$$\frac{\text{output from projector (in lumens in this case)}}{\text{target illumination intensity in lux } (= \text{lumens / sq m})} = \begin{array}{l}\text{area of screen} \\ \text{in sq metres,}\end{array}$$

so from the above:

$$\frac{4000}{400} = 10 \text{ square metres.}$$

If the slide format used above was 35 mm (i.e. with a ratio of 2:3) then the screen would be 3.89 m × 2.59 m. Beyond this size, the illumination intensity would fall to perhaps unacceptable levels, and larger format systems would have to be employed.

The equations above can be used to discover the light loss caused by increasing the beam angle of the lens, so that the throw shortens. The figures below are those given by Pani for their BP1.2 HMI projector which uses an HMI 1200 watt lamp. The required picture width is 5 m and Table 3.2 shows the effect on the illumination intensity of using narrow angle lenses from longer throws.

The equations can also be used to discover other factors, for example, if the illumination intensity is known, and additionally the throw and picture size are known, then the following will produce the required projector output (beam candle-power, BCP):

System output candela /lumens = illumination intensity in lux × throw squared

BCP= 908 × 14.35 × 14.35 = 186 977 candela.

Table 3.2 Comparison table for Pani 1.2 kw projector

Beam angle(°)	Lens (cm)	Output (cd)	Throw (m)	Illumination intensity (lux)
60	13.5	9000	3.87	600
46	18	17500	5.16	657
39	22	26500	6.31	665
31	27	39000	7.74	651
24	33	74000	9.47	825
20	40	108000	11.48	819
15	50	187000	14.35	908

(for picture width of 5 m)

The following data might be useful in the above equations:

1 lux = 0.0929 foot candles (fc);
1 foot candle = 10.76 lux;
1 foot = 0.3048 metre;
1 square foot = 0.093 square metres.

Finally, the data can also be used to judge the efficiency of projection surfaces by contrasting the illumination intensity with the luminance, in other words, the number of lumens arriving at the surface for a given area, contrasted with the number of lumens reflected from the same area, in a specified direction, the result indicates the 'gain' or otherwise, of the surface. For example, a metallized screen might reflect more light on its axis, and a gain of 2.5 would be common, i.e. for every (say) 10 lumens per unit area (i.e. square feet or square metres) which are received, then 25 are reflected.

Calculating beam angles

In the UK theatre spotlights generally include information about their beam angle in their title (i.e. Prelude 16/30 indicates a variable beam angle between 16 and 30 degrees), whereas, spotlights in the USA are referred to by their lens diameter (first figure) and focal length (second figure) in inches, thus, a 6 × 9 is a wider beam angle unit than a 6 × 22 because it has a short focal length lens. Projection lenses are usually referred to by their focal length (i.e. 50 mm etc.) although Pani usefully tend to include both focal length and beam angle. The relationship between throw, picture width and beam angle, is given by the following formula:

$$\text{tangent of half beam angle} = \frac{\text{half beam diameter.}}{\text{throw}}$$

The length of the outer edge of the beam can be found by:

$$\text{outer edge} = \frac{\text{throw}}{\text{Cosine of half beam angle}}$$

If the length of the outer edge is fed into the system output calculation (BCP) mentioned earlier, then the values at the edge of the picture can be calculated, of course, professional projectors are designed to distribute the light evenly, and thus the equations cannot always take account of the projectors' performance in this respect.

How bright should it be?

The main factors which influence the target illumination intensity of a projected picture, the illumination intensity we would desire, are the nature of the slide itself and the likely ambient light level on the performing area adjacent to the screen. As has been mentioned earlier, the amount of colour and detail within the slide has a major influence on the amount of light required to project that slide properly, the more colour and detail then the more light will be required. Additionally the more colour and detail then the larger the aperture, that is the larger the slide should be too, in order to let more light through the slide. Thus an 180 mm slide will offer more potential than a 35 mm slide of the same subject, not only because of the inherent brighter lightsource, but also because of the larger aperture through which the light has to pass. Some Pani projectors use 240 mm square slides and thus these machines are among the brightest on the market at the time of writing. Thus if the budget is tight, simple and monochromatic slides can sometimes be projected with better results from lower powered and smaller projectors than would complex richly coloured slides (Figures 3.1 and 3.2).

The ambient light level also has an influence as we have seen, thus if the production is known to require low light levels, then clearly the projections stand a better chance of showing through than if the production were to require high light levels. Part of this process is also influenced by the set, in terms of the way the stage and surrounding surfaces will reflect light. Consequently, a target illumination intensity level for one production will not necessarily do for another, the above factors have to be taken into account.

At this point it is useful to compare the target illumination intensity with that required by both statute and codes of practice for various tasks in industry. Whilst there are several bands of figures, the ones which are of most concern in this context are those to do with tasks in display work, libraries and writing. The level increases as the detail in the work gets smaller so that large items can comfortably be managed within a range of 200–500 lux, medium sized tasks can be managed within a range of 500–1000 lux and detailed work requires 1000–2000 lux. This is a useful guide when evaluating the amount of detail and colour in the slide to be projected. There is, of course, no statuary relationship between the illumination intensity required in industry and that on stage during performances, but they are, nevertheless, useful as a point of reference.

Figure 3.1 Pani BP6 Gold high performance 6 kW HMI scene projector fitted here with *f* 60 cm lens, grey scale dimming shutter and aircondition cooling system. Photo courtesy of Ludwig Pani.

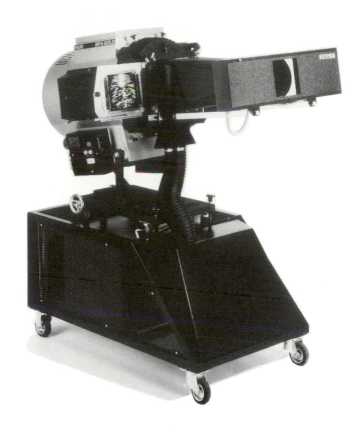

Figure 3.2 Pani BP4HMI 4 kW scene projector fitted here with *f* 13.5 cm lens and grey scale dimming shutter. Photo courtesy of Ludwig Pani.

Theatre lighting people rarely become involved with the terms 'illumination intensity' and 'lux' in their everyday process of lighting productions, but the calculations are useful when working in unusual circumstances. On these occasions it is wise to compare the figures for the new item of equipment (i.e. the projector) with figures from situations with which the person is familiar, and which were acceptable, hence the following figures for the illumination intensity produced by conventional theatre floods at average throws:

500 watt flood gives 390 lux at 3 m throw;
1000 watt flood gives 360 lux at 4 m throw.

The following are the figures for conventional theatre spotlights, each at a throw of 10 m (note that precise comparison with projectors is not possible because of the effect of the slide, the objective lens and the heat absorbing glasses which can reduce the light output of the projector by as much as 20 per cent):

650 watt 28/40 degree profile gives 360 lux;
1000 watt 50 degree profile gives 640 lux;
1200 watt 26/44 degree profile gives 1051 lux;
1000 watt Par CP62 24 × 12 degrees gives 1100 lux;
2500 watt 20/38 degree profile gives 2002 lux.

A comparison with film and TV lights also at 10 m throw, gives the following figures:

575 watt MSR or HMI 47 degree fresnel gives 252 lux;
1200 watt MSR or HMI 53 degree fresnel gives 440 lux;
2500 watt MSR or HMI 62 degree fresnel gives 980 lux;
4000 watt HMI 52 degree fresnel gives 1660 lux.

The widest beam angles have been chosen from the above equipment because, this makes it easier to compare performance with projectors which, on stage at least, would tend to be used with wide angle lenses. Thus the figures for projectors at a 10 m throw and with the widest angle lenses are as follows:

BP2 2000 W Tungsten 60 degrees lens gives 47 lux;
BP1,2 1200 W HM1 60 degree lens gives 90 lux;
BP2500 Halogen 2.5 kW Tungsten 63 degree lens gives 150 lux;
BP2,5 Compact 2.5 kW HMI 63 degree lens gives 280 lux;
BP4 Compact 4 kW HMI 63 degree lens gives 728 lux;
BP6 Gold 6 kW HMI 63 degree lens gives 920 lux;
BP12 Platinum 12 kW HMI 62 degree lens gives 2.240 lux

A final piece of the jigsaw is the transmission factor of the slide itself which obviously cannot be judged in this book since each slide will vary. Nevertheless some guidance can be found from the transmission values of theatre colour filters. The majority of filters used for general stage lighting – that is not for special effects – are likely to have transmission values in excess of 50 per cent, and many of the paler or 'no-color' filters will have transmission values in excess of 80 per cent. Naturally projection slides are likely to have many colours within them but some comparison with the theatre colour filter range will enable the prospective projectionist to form an idea as to how much light will be lost by the slide. This means that it is reasonable to suppose that as much as 50 per cent of the light of the projector will never reach the screen and this must be taken into account in the calculations.

It is useful to remember that experimenting with different exposures, when the transparency is made, can permit more light to be transmitted, and could even result in a lower powered projector being used.

An additional factor is the screen, and we have seen earlier that whilst some screens have a high reflectance value, other screens, such as those used for rear projection, will absorb a lot of light. The net result of all this is that the target illumination intensity is unlikely to be lower than 500 lux (except in special circumstances) and more likely to be closer to 1000 lux, going beyond this if the slide is highly coloured, rear projection and/or high ambient levels are involved. Finally do not forget that dust and grease will absorb a large quantity of light, so regular maintenance of the reflectors and lenses is essential.

Controlling the spill

The projected picture will not be effective if the surface onto which it is projected is already receiving light from elsewhere and part of the success of projection is to control this 'other' light. Light which hits the projection surface directly could come from exit signs, daylight, badly focused stage lights and uncontrolled work lights. In each case the problem can be identified in advance by visiting the venue and examining the relationship of the fixed light sources to the planned location of the screen. Lights which are required by the licence, such as exit signs, secondary and maintained lights must not be tampered with but sometimes baffles can be fixed to mask the spill whilst still enabling the fitting to do its job properly. Stage lights must obviously be controlled and work lights fitted with baffles and/or colour filters.

The main damage will come from the stage itself and this, in turn will come from the spill from the stage luminaires and from the reflection from the stage floor. In the first case the choice of luminaire is critical, fresnels, prism convex, and par spotlights all give more spill than profiles (lekos in the USA) and it would be the latter which would be the natural choice in projection situations. It should be noted that any spotlight will give more spill if its lens is not clean! Many profiles and all lekos can additionally be fitted with a 'tophat', a cylindrical tube which fits into the colour runners and which additionally cuts down the stray light. If 'tophats' are not available, then the proprietary 'blackwrap' from Rosco is very useful. This is a black matt aluminium sheeting which can be cut with scissors and bent to any shape to control the light further.

Since light travels in straight lines if unimpeded, the direction which the spotlights take onto the acting area influences both the direction of the spill, and of the light reflected from the stage floor. Generally the spotlights should be aimed parallel to the screen surface and, since most screens are at the rear of the stage, this means that the natural type of lighting for projection is crosslight or sidelight. This will enable the stray light and much of the direct light to pass into the wing on the side opposite to the location of the spotlight, thus keeping the screen clean of light. Of course this depends entirely upon the design of the set and this is one reason why the people responsible for the projection and the lighting design must be involved in the concept discussions for the set, the discussions which take place before the model is glued down. If there are no wings as such, then crosslight is still possible from the ends of the spotbars, a format sometimes known as an 'end pipe shot' in the USA, and indeed the format chosen by the author for his lighting of The Tales of Hoffman at the Opera Company of Boston, when the screens blocked out any normal sidelight (Figure 3.3).

The material of the stage floor is also a critical factor as it can create either diffuse or specular reflection (Figure 3.4). The former is the more desirable for projection since there is less light reflected in any one direction than in specular reflection. This means that in projection situations the stage floor should not be made of shiny material. People have gone to many lengths to create the perfectly matt stage floor and in his book *Stage Lighting* Richard Pilbrow describes one occasion in New York when the stage floor was painted with a special paint, claimed to absorb 98 per cent of the light; sadly it also appeared never to dry, with disastrous results, Mr

Figure 3.3 When two angled screens cut out the side light for *The Tales of Hoffman*, the author used the ends of the spot bars to light the singers, still minimizing the lightspill onto the screens (see also Figure 4.1).

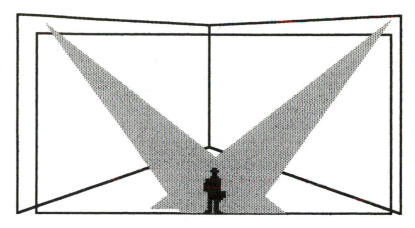

Figure 3.4 On the left a light beam is diffused by the surface, thus the intensity of the reflected light will be far less than that shown on the right, in which almost all the light is reflected at an angle equal to that by which the light arrived. This is known as specular reflection.

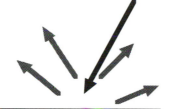

Pilbrow does indicate, however, that according to the manufacturers the problem 'has been corrected'. Certainly the textured side of hardboard is useful when painted matt black, and other materials such as carpet (without the highly inflammable backing) or black velour, have been used with success.

In musical productions when follow spots are used, their location can frequently lead to reflections on the screen. Follow spots are usually mounted on the centre line, which is exactly the wrong location for this problem but in addition, they are also the brightest light and so will reflect more from any surface.

If possible, the follow spots should therefore be moved so that their reflected light does not hit the screen. This will usually mean moving them round to the sides which provides a better angle for them in any case. Sadly this is more expensive since one spot per person is sufficient when they are located on the centre line, but two will be required from the sides.

Making slides

Before considering how slides are made, it is necessary to identify the main types which are used in projection. Some can be made by the prospective projectionist in his or her theatre, while others will need to be produced by an outside agency.

In terms of photographic slides in theatre, the main sizes are likely to be 35 mm, 180 mm, and 240 mm and it is clear that few people outside specialist photographers will possess the necessary cameras which are used to deliver the sizes beyond 35 mm on sheets or plates.

3¼" × 4¼"	83 × 121 mm	(quarterplate)
	90 × 120 mm	
4" x 5"	102 × 107 mm	
4¾" × 6½"	121 × 165 mm	(halfplate)
6½" × 8½"	165 × 216 mm	(fullplate)
8" × 10"	203 × 254 mm	
11" × 14"	280 × 356 mm	

Thus 35 mm is effectively the dividing line between what can be produced 'in-house' and what needs to be sent away to one of the specialist companies. All companies renting projection equipment offer a slide manufacturing service, either directly or through their own contacts. The advantage of this is that it puts the client in touch with a ready pool of expertise on the whole subject, and the benefits of an early contact cannot be stressed sufficiently.

Whilst undoubtedly related to projection, photography is a separate skill in itself, and thus this book is unable to cover photography in detail, the reader will undoubtedly be able to locate some of the excellent books on this subject (notably those by Michael Freeman) which are mentioned in the Bibliography at the rear of this book. Readers concerned with multi-screen, educational and display projection would also find Robert S. Simpson's book, *Effective Audio Visual* very useful. Nevertheless some general guidance can be given here.

First, film stock can vary from batch to batch, and thus it is vital that slides that are to be projected in consecutive sequences or in adjacent locations are all photographed from the same batch of film. This might be difficult if the all artwork for the slides is not available at the same time, and one solution is to use one camera specially for this purpose and no other. However the disadvantage of this is that the whole run of slides cannot be seen until the last piece of artwork is photographed and, since it is wise to test the slides well in advance of the fit-up period, it is clear that the artwork must be available in good time. Some judgement is required here about the sensitivity of the material involved and the need for continuity and care. Another factor is the speed of the film – fast film tends to be grainier than slower film, and could thus offer a poorer picture, but some slow film cannot be processed domestically and would need to be sent away. Those accustomed to working only in negatives will find that transparency film has a narrower exposure latitude. If the subject matter is only available in photographic form this can be turned into a transparency, but with some deterioration, and so original artwork or original location photographs are the best for high quality work. Slides can also be produced by drawing on theatre colour filter.

All film expands with the heat of the projector, even in those cooled by fans, and so it is vital that the 35 mm slides are mounted in hardened glass so that the film remains stationary and thus in focus, all glass for slide mounting is specially formulated to protect against the 'Newton's rings' effect, in which coloured fringes can appear when light passes through two surfaces separated by a gap which is sympathetic to the wavelength of light. For high powered projectors glass-mounted slides offer higher light transmission than film and are easier to work with when handpainted images are required, with the transparency sandwiched between two pieces. Film is more durable and more easily duplicated and should be held in a metal holder. Cibachrome P-30 or P-3 is said to provide the best heat resistance and longevity.

Slide mounting must be carried out in a clean environment, and care taken so that moisture does not enter between the film and the glass. The cardboard mounts for 35 mm transparencies should not be used. It should be noted that the mounts for all transparencies are slightly smaller than the actual frame, in order that a crisp edge can be projected and also permit easier mounting, thus whilst the image of 35 mm transparencies is 36 mm × 24 mm, the actual aperture is 35 mm × 23 mm, hence the nomenclature '35 mm'. Similarly with larger slides, the 180 mm square slides associated with Pani and other large scale projectors, actually have an aperture of 168.6 mm × 168.6 mm for glass manually changed slides and 155 mm × 155 mm for metal film holders used in automatic slidechangers. 35 mm slides are associated with multi-screen or consecutive sequences and in these situations precision mounts are required, these are available with sprocket holes which precisely locate the film, but, whilst they can be useful in many situations, they are only really effective when a special register camera has been used to locate the sprockets correctly to the film frames.

Slides also require a label which will not impede the movement within the projector, the label should allow the slide to be identified with respect to its location in the slide tray, or production sequence, and should also enable the person operating the projector to correctly appreciate which way to load the slide into the projector. In the case of a magazine type projector it is wise to put a dot in a corner which can still be viewed when the slide is loaded, so that it is easy to see which slide is incorrectly loaded. Other than to prevent light leaks, slides should not be masked on the transparency holder itself, but instead the projectors' own masking system, or some home-made alternative, should be used. This is because the positions of projectors, screens and scenery have a habit of changing at the last minute!

Overhead projection transparencies, which can be either 10" square (250 mm × 250 mm) or more recently A4, can easily be made by working on the transparency with proprietary marker pens which are available in various colours. However, a far more professional appearance is possible through photocopying printed material onto the transparency and most High Street 'Copy-shops' offer this facility. It should be noted that typewriters and dot-matrix printers are not suitable as the quality of the image is frequently poor and the print is likely to rub off; laser printers offer superior quality and fastness of the image. Note: while many laser printers, bubble jet and photocopying machines can work to acetates, not all acetate is applicable because the heat of the laser/copier causes it to melt in the machine. Check that only special heat resistant OHP acetates are used. Computer generated text and artwork is now within the reach of most people and among the best is the Apple Macintosh range, one of which is being used to write this book. See also overhead projection in Chapter Two for information on LCD screens.

Gobos can either be made 'in-house' or purchased from one of the main suppliers who will not only have a wide range of standard patterns, but who will also offer a specialist service of making gobos to the client's own artwork. As with the production of slides, this contact enables the client to tap into a range of advisory and rental services. The timescale on making gobos is usually about 10 days and therefore the artwork needs to be available in sufficient time before the fit-up period. Standard patterns are not always in stock and so again some notice is wise. For those gobos made in-house the most suitable material is printer's lithoplate which can easily be cut with sharp scissors or a knife. This process does not, however, lend itself readily to fine detail, and so those with access to some metalworking or engraving facility are more likely to produce sophisticated work. Some domestic baking and freezer containers can also be used for very basic shapes,

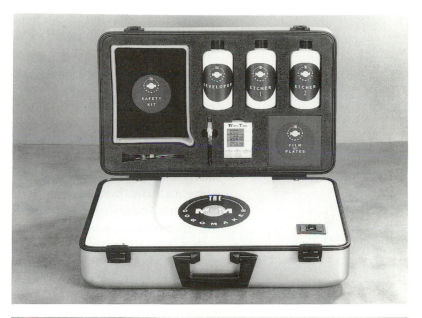

Figure 3.5 Gobo-making kit from M&M. Photo courtesy of M&M.

but many are coated with a film which, being inflammable, must be removed before use. Most gobos produced in house are likely to have very sharp edges and care is needed in their handling unless the edges can be filed smooth. It is also worth noting that there is a wide range of expanded metal grilles which can also be used to project regular shapes. There is also now a kit which enables venues to make their own gobos by etching their pattern onto the aluminium provided, the manufacturers claim that this is stronger than the stainless steel of most conventional gobos and the method is cost-effective for venues which require a lot of special shapes or symbols (Figure 3.5). All slides, gobos and OHP transparences should have back-up copies toegther with any relevant special glass and mounts.

Size of lettering

One of the most frequent mistakes in both overhead projection and in slide projection of text is to make the size of the lettering too small to be read adequately and easily by the audience. In overhead projection especially, the tendency is to take some page from a report, and simply photocopy it onto the appropriate acetate sheet. First it is highly unlikely that text written on a normal A4 page would be large enough and, secondly, the size of the letters would also tend to mean that they were thin and lost on an illuminated surface. The same would apply if the page were to be photographed and mounted as a slide.

The author's rule about the size of projected letters, on overhead transparencies, is that they should be at least one-twentieth the size of the transparency height; assuming the picture will be 2 m high, this will produce projected letters about 100 mm high; other projection experts such as John Simpson (see Bibliography) suggest no more than 15 lines of text per transparency with a good space in between. For slides, the lettering needs to be much bigger because generally OHP will be used in smaller theatres than would slide projection, and the audience will be further away. Bold lettering and a simple sans serif typeface can help. If it is not possible to print the original artwork via computer or rub-on lettering, then it is wiser to hand print rather than to type. There is also a view that for theatre work, if the line thickness is sufficient, it is better to have a reversal so that white letters (or graphics) show against a black (unlit) background.

Surtitles

This is the term applied to the translations of libretti in opera which are usually projected either above or at the sides of the proscenium arch. Their use is controversial because purists feel that it is a distraction, and this is certainly true if it is not done well. The best surtitles in the author's experience were those developed by Lisi Oliver for the Opera Company of Boston in the USA, Ms Oliver has the advantage of being an opera director so, being knowledgeable about the productions, she usually cued the changes from one line to another sympathetically. In addition Ms Oliver is a considerable linguist, and thus at home in the many languages in which opera is sung – many of the surtitled productions were her own translations. Such combined talents are rare, but certainly the person responsible must have some understanding of the production. It is best if the screen is above the stage rather than at the sides, because this latter position can be distracting. Patrons need to be told when booking seats if the screen is not visible from, for example, rear stalls because of the overhang of the circle. Additionally the design of the screen needs to reflect the design of the production. The slides should be white lettering on black and should crossfade from one to another (Plate 3.1).

Copyright

In the enthusiasm to mount the production, the ownership of material should not be overlooked. Generally the person who produces the material will own the copyright, so the designer will own the copyright of his or her artwork, the photographer will own the photographs or slides. Commercial photocopying outlets are not supposed to work with material which is clearly someone else's copyright, such as photographs from a published book. In all these cases permissions should be sought and relevant people credited in the programme.

Distortion

Distortion will result from any kind of projection if the projector is not square to the surface on which the image is being projected. However, the distortion is not necessarily a problem in all cases, because the projection of abstract shapes (such as 'breakups' from gobos for example) could benefit from the

Figure 3.6 Modelbox computer aided lighting design program here showing the ellipse of light beams from specified positions. Photo courtesy of Modelbox.

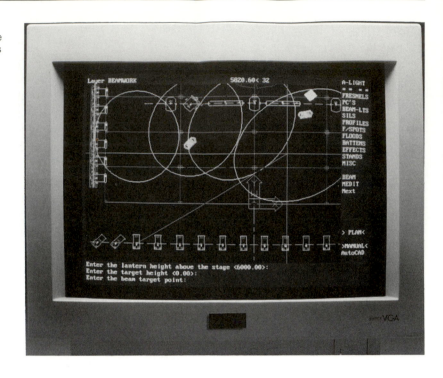

distortion if it introduces an additional dynamic to the overall result. Some computer lighting design programs (such as those offered by ModelBox in London (Figure 3.6)) can demonstrate the amount of distortion present from a projected gobo at a defined angle. Furthermore there is a cost factor involved in overcoming the distortion, and this should be measured against the gain involved. Abstract patterns might not require correction but geometric shapes would because the distortion would be more apparent.

The simplest method of overcoming distortion is to move the screen so that it is square to the centre line of the projection beam. This is the technique employed in most large cinemas where the projection room is usually at the rear of the circle and thus the projection beam is at an angle to the vertical, resulting in the keystone distortion mentioned earlier in the book. On stage, if the screen is framed, it is an easy matter to arrange the angling process; many people will be familiar with the angled screen associated with overhead projection which is done for the same reason.

If this is not possible, and if the projector cannot be relocated (say as in back projection – see Chapter Two), then the choice of projector should be checked, because, some 35 mm machines

Figure 3.7 The camera must
photograph the art work (left) at the
same angle, and with the same
focal length lens, as the projector
will make with the screen on stage
(right).

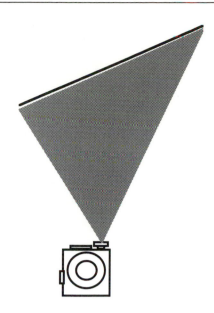

Figure 3.8 Artwork made into a
predistorted slide by the students
of Croydon College England for
their production of *Cavalcade*. This
was done to overcome keystone
distortion

can be provided with lenses which will angle so as to correct
small amounts of keystone distortion. Otherwise special slides,
pre-distorted, will have to be made.

Such special slides are easier to make than many people
imagine. The artwork must be photographed from the same
angle and relative throw (to picture width), i.e. with the same
focal length lens as the projector will employ on stage. The
resultant slide is distorted, but since the projector will reverse
this, it will project an undistorted picture if placed in the
projector in the normal way (see Figures 3.7 and 3.8).

One advantage of using 35 mm is that the slides can be taken
and developed domestically or locally, and with a quick
turnaround, so that it is possible to see if the pre-distortion has
worked by either projecting in the actual environment or on a
scale model in a room. Adjustments can then be made, and
another slide taken. This would not be impossible with larger
formats but, since the slides would have to go away to be made,
it would tend to discourage this kind of experimentation.

Outdoor events

The principle potential problems of most outdoor events are:

- the interference of natural daylight;
- the damage done to equipment by the weather;
- the lack of adequate, or any, power supply.

Natural daylight

Lighting-up times relevant to the location can easily be obtained from local reference libraries, the police or motoring organizations, and sunset times are also available in the UK from the Meteorological Office in Bracknell and the Royal Observatory. The latter will also provide information on sunset times anywhere in the world. These figures should be used as a guide only, because total darkness does generally not occur until sometime later, on a clear evening perhaps as much as an hour later. However, by sunset there should be no direct rays of sunlight to wash out any projection, even if there is still an afterglow in the sky. Assuming that the event is planned far enough ahead, a 'matching visit' can be made because of the earth's movement around the sun, the sunset time on the day of the performance occurs on another day in the year, thus it is possible to visit the site in say spring, and find out exactly what the ambient light level might be in the autumn (and vice versa). Naturally the weather will be a contributory factor in terms of likely cloud cover and so on, and again the Meteorological Office can furnish figures, local airfields should also have this information.

The matching visit, which must be made at the time planned for the performance, will also provide information on the spill cast by surrounding streetlights, advertising signs and neighbouring buildings, a factor which it would be difficult to estimate in daytime. Obviously lights which are required for safety and security, such as street lights, cannot be turned off but many local authorities are amenable for baffles to be erected in order that the spill can be contained. This needs to be demonstrated to the representatives of the Local (licensing) Authority in sufficient time for them to make the necessary official arrangements. Neighbouring lights might also prove a problem in that they appear within the vista of the event, and create a distraction or distortion of the overall design, even if they do not provide spill onto the location itself. Under these circumstances if the erection of baffles is not possible, then the re-orientation of the vista should be considered.

The Ancient Greeks oriented their outdoor theatres specifically to take advantage of the rays of the sun, by placing many theatres so that the sun was behind the audience, which would thus not dazzle them but illuminate the actors. Other theatres did face westerly so that the setting sun provided a beautiful backdrop behind the open stage. This is the author's experience at the Minack Theatre in Cornwall, which is hewn from the rock close to Land's End; creating beautiful lighting artificially is dispiriting when you have to compete with rather superior sunsets, which every member of the audience can clearly see! This suggests that the location of the sunset should be taken into account if it could unduly dominate the scene.

Although not the first, or indeed the only exponent of outdoor projection (see Chapter One), Jean Michel Jarre has become synonymous with this technique through his world-wide outdoor concerts which began in 1979 in Paris in front of 1 million people. Peking and Shanghai followed in 1981 and Houston in 1986, where the crowd grew to 1.5 million, then Lyon in 1987, and London Docklands in 1988. Many smaller (a relative term!) concerts have also been given, most recently at the Mont St Michel in France, and a concert at Wembley Stadium, London is planned as this book is being written (Plates 3.2 and 3.3).

Jarre's music is specially written to provide opportunities for the triggering of visual effects, and the concerts feature light and laser shows, fireworks, film and still projection. A significant aspect of the work is the linking of all these items to central computer control through time codes, even the operation of the fireworks, which is considered a 'first'.

Jarre uses Pani and 'Hardware for Xenon' projectors, and his concerts have provided excellent showcases for these companies' products; in turn many refinements have been developed specially for the Jarre effects. In fact, the projection aspect of the concerts are fairly routine by comparison with the logistics of providing sound to 1 million people, and rigging and controlling the other elements. Power generation, too, is immense with between 1 and 2 megawatts being common. The range of projection units used includes Pani BP6 Golds, often with 800 mm lenses for the very long throws on the main images, and many smaller HMI 2.5 kW units, with rotating effects wheels, for peripheral areas and routes to the sites. Jarre's use of the slide changers is also interesting in that he tends to use the actual effect of the slide moving, either vertically or horizontally, as a visual component of the show, this is by contrast to normal use where the slide would usually be changed once the mechanical dimmer has shut out the light (Figure 3.9).

The timing of the Jarre concerts has occasionally given cause for criticism, notably the London Docklands event which was performed in October, a month not noted for its good weather, and consequently many people left during the

Figure 3.9 Hardware for Xenon's 5000 DHX projector for 18 3 18.5 cm slides with automatic changer. Photo courtesy of Hardware for Xenon.

evening as the rain and cold took effect. High winds also blew over a followspot, and the cold air acting on the hot HMI lamps caused many to blow. A reminder that, however much protection is provided, nature will always find a way.

Weather damage

In addition to the circumstances mentioned above, damage would usually come from rainwater, dew or condensation, and each requires specific protection. In the first place some hire companies will not permit their equipment to operate outdoors at all, and this should be checked otherwise infringements of the hire company's terms of business could cause the insurance to become invalidated. Hire companies are also useful centres of advice on which machines are more likely to withstand a degree of exposure.

The location of the projectors will obviously have a bearing on the kind of protection which can be provided and, since projectors are expensive, there is also a security aspect here. If at all possible, the projectors should be located within a structure, a local building, hired Portakabin, or specially built booth. At the very least, the projectors should be provided with overhead protection from rain, and this needs to cover the whole area surrounding the projectors to protect the operators and to provide additional cover against rain driven by the wind. Polythene sheeting can then be attached at the sides to provide further protection. Some projection areas will require heating at night to prevent condensation.

Power outdoors

Few outdoor locations have sufficient power to feed a battery of projectors, and the related event and site lighting required by the performances. Even locations which already have power, because they have another normal daytime function, might not have sufficient power as it may not be possible to divert that power for night-time performances if some industrial process has to be maintained. Thus, the first piece of information to be gathered is the location of power supplies, and their availability not only on the performance nights, but also on previous nights for focusing, plotting light levels and rehearsing. It is worth noting that the owners or managers of many locations do not know what power they have at their disposal, because people only use the power they need and are unaware of what is available. The local Electricity Supply company will be able to provide plans of the location showing the power intake, and the route that its distribution takes around the area. This might uncover an unused power supply in an old building nearby which can be accessed for the event. Certainly the Electricity Supply company and Local (licensing) Authority should be consulted about the arrangements, even if there is adequate power available.

Cable runs should be well planned in advance, not only for the obvious reason that large power cables are expensive to hire, but also because the Local (licensing) Authority is unlikely to approve of such cables coming close to members of the public, and will certainly not approve of the public walking across them. This suggests that bridges, cateneries or ducts have to be constructed and these items are often overlooked both in the budget and in the estimate of how long the fit-up will take.

Long cable runs will cause voltage drops, and this can be calculated from data supplied by the cable manufacturers, which indicates voltage drop per amp per metre. Once the calculation has been done the answer is expressed in millivolts which should then be divided by 1000 to give the result in volts. Voltage drops on their own are not likely to cause a serious problem if the projectors are fitted with incandescent lightsources, beyond of course reducing the light level.

However, voltage drops will cause problems if the lightsource requires a specific voltage to fire or maintain its charge, and many projectors require additional power for fan cooling, slide change, or effects motors which will not operate below a specified reduction in the voltage. Under these circumstances generators may have to be used, even if there is actually sufficient power on site overall, but it is in the wrong location and too far away.

Generators are rated in Kilovolt amps, abbreviated to kVA, and this measurement takes account of the kind of loading to be attached, each kind of loading having a factor number which influences the simple product of the multiplication of the voltage by the current. The kind of lighting usually associated with these events has a factor of 1, so that the product of the voltage and the current is unaffected, and also matches the power rating in watts. Thus five 5000 watt incandescent projectors providing 25 000 watts or 25 kW of power, and which would draw a total current of 104.16 amps at 240 volts, would require a 25 kVA generator. Generators produce noise when in operation, and even the so-called 'silent' types still produce the noise equivalent to a car engine, thus the siting of these devices is critical if the event (and the neighbours!) are not to be disturbed. Unless the generator can be placed behind some structure which can mask the sound, some sound baffle will have to be constructed. Often the natural landscape can also help, parking the generator behind the nearest hill for example. Whilst generators can be small, the types associated with outdoor events are likely to be large, and mounted in vehicles, thus the site must be capable of accepting the weight of the vehicle – not a problem on an urban or industrial location, but a real problem in a muddy field! Generators require fuel, and if the performances run for many weeks, which might be the case in say summer *son et lumière* events, then additional fuel will be required. It might be easier to bring this to site (rather than move the generator) but access for this will have to be planned in advance and, of course, the cost included in the budget. The generator hire company will be able to provide an estimate of the fuel required, once they know the electrical load and the times it will be needed.

It is important to remember that the early estimates, before the event, on which the hire company quoted, might become unrealistic if the later rehearsals make changes to the production, and so a contingency in the budget is vital. Generators also come with an engineer, and this person will charge a daily rate and expenses which could include the cost of a hotel if

Figure 3.10 Schematic site diagram by the author for outdoor event showing power distribution.

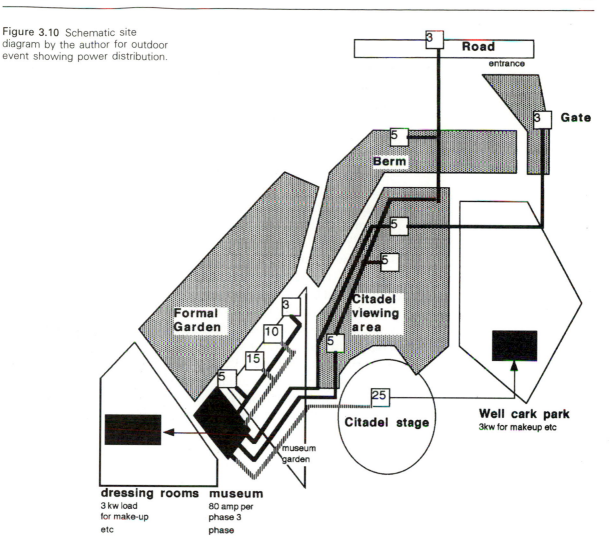

Area	time	phase 1	2	3
Formal Garden	6.30 - 8.30	8 kw	-	-
"	8.30 - 10.00	4 kw	-	-
"	6.30 - 8.45	-	25 kw	-
Citadel *	8.45 - 10.00	-	25 kw	-
Citadel Viewing	7.30 - 10.30	-	-	10 kw
Berm downlight	6.30 - 10.30	5 kw	-	-
Berm specials	6.30 - 9.00	-	-	5 kw
Gate and bridge	6.30 - 10.30	3 kw	-	-
Road	6.30 - 10.30	-	-	3 kw
well dressing room	6.30 - 10.30	-	3 kw	-
museum dressing	6.30 - 10.30	3 kw	-	-
(security light	*10.00 - 4 am*	*2 kw*	*2 kw*	*2 kw)*
total connected loads		19 kw	30 kw *	18 kw

* the 25kw in the Citadel Stage will be dimmed and thus will not draw full power.
NB Each area will also have local security light on the same phase, total load 6kw

figure in box represents no of kw distributed to that area, and thus no of circuits then distributed locally. via mcb

the location is far from the engineer's base. Many engineers will, however, help with the general site work, and thus provide extra value for their client, but this aspect should be checked with the hire company first. It is wise to distribute the overall site power to several sources in case of failure, and thus if no power is available on the location at all, then more than one generator should be hired, one for the production lighting, and one for the site lighting and sound. The separation of site lighting might have to be taken further, and backed with battery exit lights, and other maintained route signs and lighting, as a condition of the licence. These items will all increase the ambient light and interfere with the vista as mentioned above (Figure 3.10).

All power supplies which do not come from the generator will have to be provided with a Residual Current Device (RCD) which will cut off the power if a fault is detected. It is illegal in the UK to fit RCDs in such a way that the loss of power causes total loss of illumination in an area. Thus it is usual to fit each circuit with an RCD, so that the loss of, say, one projector, will not affect the others. RCDs need to be located so that they can be reset after the fault has been corrected, and this means that they should not be buried under masses of cable, or in some power switchroom to which access is difficult. All plugs and sockets should be weatherproof outdoor types, and not those normally associated with indoor stage lighting. Where indoor types are unavoidable, as they would be on the projector itself, then suitable weatherproof protection should be provided.

Hired equipment should come with proof that it has passed a test such as that with a Portable Appliance Tester (PAT), to satisfy the requirements in the UK of the Electricity at Work Act. This test covers the devices' insulation resistance, and includes a visual check of the condition of cables and connectors. Long temporary cables runs will only pass the test if they are provided with conductors of the correct cross-section.

Summary

Whilst it is perfectly possible to produce excellent pictures through trial and error (and it is a good exercise for students to attempt to do so) almost all aspects of the projection process can be calculated, notably the picture size, beam angle of the projector, light output of the projector and resulting amount of light falling onto the projection surface. In addition to the performance of the projector and screen, the final intensity of the picture is influenced by the amount of colour and detail contained within the slide, and by the amount of extra light present in the neighbourhood of the projection surface and thus washing out the picture.

Slides which are to be projected in sequence need to be photographed from the same film in order to overcome variations in film stock, and the slides need to be mounted in glass to prevent movement of the film whilst in the gate of the projector. Most slides beyond the 35 mm format will require special attention by professional studios and will not be made in-house. Overhead transparencies, however, can be produced through word-processors and professionally photocopied onto the proprietary acetate sheets required. Lettering on slides and transparencies requires special attention in order that it can be legible from long distances, and this usually means that the lettering has to be larger than is commonly supposed. This is particularly important if the text is a key component in the production as it is with the projection of translations of operas. Predistorting the slides during the photographic process can help to overcome distortion, although angling to projection surface, and choosing distortion-correcting lenses can also help.

The use of projection outdoors requires greater control over the ambient light, part of which will involve an identification of typical sunset times and neighbouring streetlighting. Weather damage also has to be considered, and this would include the effects of rain and wind in addition to temperature changes. Power generation and cable runs need to take into account the requirements of the licensing authority.

4
Scenic projection

Who does what?

When projection is used as a component of any performance, it is important that both the correct hierarchy is established between the members of the creative team, and also that the production schedule and budget take account of the needs of the medium. All too often, projection is not viewed as an additional department in the same way as costume, set, sound and lighting which each have their own team and budget. Perhaps, this is because the realization of projection involves more than one department, and thus there is a view that the workload can be shared. In most uses, the responsibility for acquiring, rigging, focusing, and operating the projectors will fall within the lighting department, the work will be carried out by the chief electrician and his or her team. However, the artwork for the slides will tend to be the responsibility of the set designer. Whilst this division of responsibility does have its own logic, it ignores the fact that there are few productions on which people are not heavily engaged, and good projection cannot be delivered by people already working at full capacity on other parts of the production. At the very least, a separate technician should be engaged, specifically to look after the projection, and interface between the electrical and design departments; he or she can liaise with the companies providing the equipment and the slides, and will probably be the operator of any special slide change, dimming or effects devices used in the production. Part of this appointment will also be informed by an assessment of the skills of the existing staff, not everyone is as skilled in the techniques of projection or equally familiar with all the relevant equipment. It therefore makes sense to hire someone who is, and this also provides an opportunity for the 'expert' to train the other members of the electrics team in the equipment involved.

A reality which also informs this matter is that few theatres will possess projection equipment, even on a small scale, and therefore, it will have to be hired. Equally, few theatres will have the facilities in house to photograph and produce slides, and thus outside companies will be consulted to provide these items. The instant this is done, the theatre is then plugged in to a fund of advice and expertise on which it is possible to draw.

Timing

It is a well-known problem, that many lighting designers tend not to be engaged sufficiently early for them to be involved in the conceptual discussions about the production which take place between the director and the set designer. For example the author has been involved in such discussions in fewer than 5 per cent of the 500 productions world wide which he has lit, and the author's arbiter is a preference to be involved before the set model is glued down to its base-board! In most situations the lighting designer is presented with a model which will already carry with it a presumption about the general style of the production, and the access which the lighting equipment and resulting light beams can have with the area to be occupied by the performers. Other presumptions may also have been made about what the lighting can contribute in the absence of adequate stage space, set budget, or even people. When projection is brought into this situation, the probability of delivering good projection declines sharply. For example, the author was once engaged to light a musical in a regional producing theatre. The decision had already been taken to use scenic projection, because most of the scenes were short and took place on large locations. The company had limited resources and could not have afforded to build or change realistic scenery. In itself, this decision was logical, but it ignored the fact that the company could also not afford to hire sufficiently powerful scenic projectors for the scale of the projection which had been decided upon before the author was engaged. With three weeks to opening night, there was little alternative but to proceed as already decided, and the final result bore testimony to the flaws in the decision-making process since the projections were not as bright as they should have been. Wiser and earlier counsels could, perhaps, have suggested one small screen, instead of three large ones, but not when the screens were already made and the production was already well into rehearsal!

The set design cannot proceed until the decision to use projection is thoroughly investigated, and this process must involve staffing, finance, timing, space and location. The location of the projectors is obviously critical, and not a factor about which there can be much negotiation, since light travels in straight lines if unimpeded, if you want the screen *there* then the projector has to go either *here* or *here*. If the location is considered early enough, then the location can be prepared and the consequences assimilated. For example the author was once engaged to design the lighting for *The Tales of Hoffman* for the Opera Company of Boston in the USA (see Chapter Three). The Company had decided to hire a set for this production because of the demands upon it made by the rest of its repertoire (which the author was also lighting). Two sets were considered, one of which used back projection. However, since one projector could not be located within the stage, where it should be if its beam was to deliver the correct image, the production manager had, quite logically, indicated a preference for the other set. This reckoned without the resources of the Company's Artistic Director, Sarah Caldwell, who simply suggested that a hole should be made in the theatre's back wall to accommodate the projection beam. When told that this would still mean that the projector would be in the middle of the street outside, she simply retorted 'We own the street, close it!' Thus it was that a valuable Pani projector was daily trundled out into the street, and covered with a weatherproof construction (christened the 'Pizza Hut' by the crew) and guarded by one of Boston's armed police (Figure 4.1).

Projectors require some structure on which to stand, and this must also be taken into account in the planning process, as it will need to be budgeted for and built, and occasionally if the projectors are to be suspended, then some structural advice may have to be sought. For example – and once again for the Opera Company of Boston – special effects designer Esquire Jauchem, sought advice on the safe working load of the dress circle front at Boston before he authorized the rigging of a battery of projectors on the rail.

It is wise to schedule the projections to be ready for two weeks before the production moves into the theatre. This gap

Figure 4.1 Sketch plan of the stage at Boston Opera for *Tales of Hoffman* showing the stage left projector in the street projecting through the access doors in the back wall.

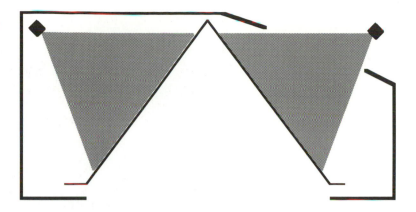

should take account of any slippage in the production of the slides, which depend initially upon the designer producing the artwork at what is also the heaviest time for them, as they are constantly being consulted on paint finishes, costume fittings and a selection of props. Once some slides are available, then a test should take place, if the theatre itself (or the projectors) are not available, then some simulation is essential. At the very least, the slides could be taken to the hire warehouse and projected there, or to another theatre where the projectors are located. Of course, if the projection is not full-size, then the intensity will not be the same as it will on the stage, but there are frequently other factors to test, such as distortion.

It was this thinking which led the Royal Opera House (ROH) to construct a large-scale model of the set for the *Knot Garden* in 1970. The production was directed by Sir Peter Hall who had written (in the Strand Lighting publication *Tabs*) that he had always avoided the use of projection in the past, because, the colours tended to be pale and did not permit strong stage lighting. Timothy O'Brien and Tazeena Firth's set consisted of ropes, hung vertically and with 500 mm to 50 mm spaces between each, the ropes were also made to interact with metal tubes which rose from concentric revolves, thus providing a multitude of surfaces on which to project the still and film images. The lighting was designed by John Bury. The usual 1:24 model was insufficient to test the effect of the projections on the ropes which were only 15 mm thick, and so a quarter size model was built (Figure 4.2).

The tests proved that a centre projecting position was crucial, in order to avoid not only distortion, but to ensure that

Figure 4.2 Set model used to
check the projections for *The Knot
Garden*, Royal Opera House

the shadow of each rope was thrown upstage and blended in,
it was especially important to avoid the projection on the floor
if the image of trees was to be maintained, and thus a low front
position was essential. Six Rank Strand 4 kW Pattern 752
scenic projectors were placed on the bridge, behind the proscenium, with slides designed to blend the image in the offstage
areas which the centre projection position could not reach. A
particular feature of this production was the specially
constructed soundproof projection room at the rear centre of
the Grand Tier, which involved removing thirty seats and
installing air handling and power systems; as if this were not
enough, the whole installation had to be capable of being
removed, and installed, as the repertoire demanded, a particular credit to the ROH team, IES Projects, the Greater
London Council (GLC) and Fire Brigade engineers and
inspectors.

The slide projectors were specially constructed by Strand
Lighting, and it is interesting to read two descriptions of them.
First, from William Bundy, then Technical Director of the
Royal Opera House;

'Mac' of Rank Strand [as Strand Lighting was then called]
. .used two 4.5 kW film projector lamp-housings as the light
sources, with his own-design optical system, a 6" Dallmeyer
lens as the objective lens and the optical faders from the

Royal Opera House Xenon follow-spots to control the projectors. The slide used was a 4.5" × 3.5". Because of the heat on the slide the colour positives had to be deposited on heat-resistant glass and the film base dissolved away, leaving just the film dyes. No cover glass could be used, and if the light source was a fraction out of alignment the image immediately burned away.

Then from 'Mac' himself, R. A. McKenzie, who played a large role in the optical systems of Strand luminaires:

Tests had already shown that the standard Rank Strand Pattern 752 4 kW scene projectors were inadequate, so a Cinemeccanica X4000 lantern with a 2.5 kw xenon lamp was mounted up with spreader and condenser lenses and a Dallmeyer 8" focus f/2 'Super Six' objective lens in the laboratory at Brentford. Experiments with this showed that satisfactory projection of 4" × 3.25" slides was possible, but that it would be necessary to use 4 kw xenon lamps if adequate intensity was to be obtained. The use of these lamps, however, made the problem of adequately cooling the slides much more difficult. Filtering out the infra-red energy was not sufficient, since so much visible light was being converted into heat by the slide. The problem was finally resolved by using glass colour slides without cover glasses and directing a stream of cooling air on both sides. In addition to this, two heat absorbing glass filters were placed in the beam between the mirror and the condenser system. The problem of handling the slides was overcome by increasing the glass size to 4" × 4", leaving a plain ¾" section at the bottom of the picture which in practice becomes the top of the slide. Special slide carriers were designed and produced to grip the sides and leave the top and bottom unobstructed. Adjustable masks were fitted to confine the picture within the stage frame.

The projected picture was measured and each projector delivered 130 lux over a picture size 68 × 41ft at a distance of 140ft (readings being taken with the slide in the projector). The film projections were apparently unsuccessful and were cut after the first rehearsal, nevertheless the final result was stunning and marked a turning point for scene projection in that it proved a perfect marriage between set and light could be achieved with careful planning and lots of imagination (Plate 4.1).

Gobos

A gobo (or 'pattern' in the USA) is the name given to a slide which can be inserted into the gate of a profile spotlight (or 'leko' in the USA) in order to project the shape on the slide. Unlike specially designed optical projectors, theatre spotlights do not contain cooling systems, and so the temperature at the gate is immense, for this reason the 'slide' is usually metal (actually stainless steel or aluminium), although heat resistant glass is also available for special images. Almost all profile spotlights contain a 'gate' into which the gobo can be inserted, but some lekos in the USA do not, and so if gobo projection is planned, it is wise to check with the rental house or theatre in advance, to see if the correct lekos are available.

Metal gobos are available in a wide variety of proprietary patterns and symbols, with foliage and windows being the most commonly used. It is possible to manufacture gobos if some basic metalworking facilities are available and certainly printer's lithoplate will suffice for basic or abstract shapes (the latter known as 'break-ups' because they cause the light beam to break into smaller pieces and add texture to scenery). Some gobos are fitted with a mesh so that finer patterns can be achieved. Expanded metal grilles are also useful as mentioned earlier (Plates 4.2 and 4.3).

Gobos themselves are circular and are mounted in holders which are rectangular. The advent of the 'tadpole' has made it possible to mount the gobo within the tadpole, so that it can be adjusted with the aid of the tadpole's 'tail' which protrudes from the gate. This avoids many burned fingers, since the chances of the gobos being correctly aligned the first time are rather slim (Figure 4.3).

Glass gobos enable sophisticated patterns to be projected, or those where the fine supporting lines of the metal structure would be too obtrusive. The opaque part of the gobo is actually chromed and reflects the light. Glass gobos require their own special holders which have wider supporting lugs for the gobo than those for metal types because the glass is thicker than the metal, and needs to expand. Additionally, spotlights which are used for glass gobos need to be tuned, so that the beam is perfectly 'flat' or even, otherwise if the beam was 'peaked' (so that the intensity increased towards the centre of the beam) then the heat would also increase, and cause the glass to shatter. In any event it is always wise to have spare copies of glass and special metal gobos – even the metal types will burn away eventually, especially if the pattern is very fine (Figure 4.4).

Figure 4.3 Adjustable gobo holder (called 'tadpole' by DHA Lighting) enabling gobos to be angled whilst still mounted within the spotlight. Photo courtesy of Great American Market.

Gobos can be used with the focus very sharp, so that the pattern is precise and clear, or with the spotlight out of focus, so that the pattern is imprecise and unclear. In the latter case the gobo will probably be used to add texture to the scenery, or suggest the dapples of sunlight through leaves. Break-ups like this benefit from broken or split colour, where several pieces are placed side by side in the colour frame. Occasionally gobos which need to be clear and sharp are foggy, and seem to have a mist around the image itself. This is a product of the type of gobo and the optical system of the spotlight, some being better than others at projecting fine shapes and lines. The addition of a 'doughnut' into the colour runners will clean up the fog, the 'doughnut' is a metal mask with a small hole punched through its centre, the hole needs to be adjusted for size because whilst small it offers the greatest clean up, but the least light, and so a balance between the two must be sought (Figure 4.5).

The rules of projection still apply, that is, the ambient and direct light has to be controlled, and the angle at which the projection beam strikes the projection surface will govern the amount of distortion present.

Gobos can also be made to order, and this is quite often the cheapest form of scenic projection since most theatres possess some profiles or lekos whilst special optical projectors almost always have to be rented, and even if the profiles or lekos have to be rented the cost is less than that of the specialist optical

Figure 4.4 DHA gobo #215, one of many standard patterns. Photo courtesy of DHA Lighting.

Figure 4.5 Half-tone glass gobo
from DHA with a doughnut used to
enhance image quality. Photo
courtesy of DHA Lighting.

projectors. The size of the gobo image cannot be simply calculated from the data produced by the spotlight manufacturer
because that is governed by the aperture of the spotlight's gate
and thus the narrower aperture of the gobo itself has to be
taken into account (see Table 4.1). Today profile spotlights are
available with considerably wide beam angles, the Turbo
Silhouette from CCT is believed to provide the widest angle,
at 58 degrees (variable from 28).

Every lighting designer uses gobos, but despite this they still
retain their impact. The author has used gobo projection on a
number of occasions, notably for *Little Tramp* described later,

Table 4.1

Gobo size ref	Overall size (mm)	Aperture diameter (mm)
A	100	75
B	86	64.5
C	150	112.5
D	53.3	40
E	37.5	28.13
G	66	60 or 49
M	66	49.5
VL1	25	18.75
J	38.8	25

NOTE: aperture here is the available size of the pattern

and for a revue in Torquay, in which the various scenes were each suggested by a gobo chosen for its suitability for the scene, a sketch about Paris was served by the Eiffel Tower, one in a cloakroom was served by brickwork, and a musical number by projecting stars. In each case the only location for the spotlight was with the lighting operator at the rear of the stalls, and so the operator changed the gobos between scenes by means of a pair of pliers to overcome the heat. The spotlight was specially mounted to limit movement and colours could also be changed as required. This was a simple but very cost-effective way of suggesting different themes, and one which served the fast-moving revue well. On other occasions, special gobos have been made for the author by DHA Lighting in the UK for a number of productions: a flying dragon appearing in the sky in *The Voyage of the Dawn Treader* was achieved through three gobos crossfading, a cottage in *James and the Giant Peach* was achieved through the projecting of the designer's sketch onto a front gauze which served for other occasions, and which, tests had proved, would not reveal other parts of the set upstage of the gauze as the projection beam passed through the gauze's holes. Elsewhere various messages were projected by gobos onto the set of *Magic Flute* as the trials by fire and water began (Plate 4.4).

Case studies

When a production is being planned which requires scenic projection, there is a fine balance between what is technically feasible from standard components, and what can be achieved through greater expenditure. Thus, as has been said earlier, the consideration about the use of projection must be made in sufficient time to investigate all the options. The following example might serve to illustrate the point.

Consider a proscenium stage with an opening 12 m wide and behind it a stage 20 m wide and 9 m deep. Imagine that the production needs to project a number of backgrounds as large as is feasible. One of the first considerations is the identification of how much stage depth the production requires. In this case if the production requires almost all the 9 m available, then clearly back projection is out of the question. If, however, less than the 9 m is required, then back projection could be a possibility. Another consideration concerns the movement of people, and especially of scenery, during the occasions when projection is required; it would generally not be acceptable for example to front-project whilst scenery is being flown out, because of the confusion of images that would result, unless of course that was the precise effect required!

Thus it is important that the projectionist is consulted before the set is too far established, and at this time the lighting designer will need to have an input so that he or she can comment on the implications on the locations of the components of the set, especially of the masking.

In this example let us suppose that the director feels he or she can accept a stage depth of 7 m; this therefore frees 2 m of space upstage which can now be considered for back projection. It is clear without even doing any calculations that there is no projection system which can deliver a picture say 10 m wide from within a depth of only 2 m and therefore the picture will have to be broken up into sections, each served by a separate projector, and the resulting images joined together to make the whole. To some extent the choice of projector at this point is influenced by the quantity of slides involved, the Kodak Carousel or Ektapro (and derivatives) offering 80 slides and the Pani range offering 15 (although special larger capacity slide changers have been made for some productions and are now becoming available generally) (Plate 4.5).

The calculations demonstrate that with the Carousel fitted with a 35 mm lens a picture 10 m wide can be achieved from 5 projectors side by side, since this picture would only be 1.3 m high, two additional rows of 5 would be required to increase the height of the picture to 4 m. Should this picture need to be faded out, and another faded in simultaneously, then twice the number of projectors (a total of 30) would be involved. It now becomes clear that additional factors have to be taken into account, a structure needs to be built to house 30 projectors, and possibly the structure needs to accommodate the fan noise, as it did in almost identical circumstances for *I and Albert* as mentioned below. If a Pani BP2 is considered as an alternative to the Carousel, then, because the Pani can be fitted with a wider angle lens than the one with which the Carousel is usually supplied (although special extra wide lenses for the Carousel are possible, but expensive) only 4 BP2s are needed for one level of the picture; furthermore because the BP2's standard slide is square (as opposed to specials for Carousels, etc.), it would be possible to manage with only 2 rows of BP2s giving a picture height of just under 5 m (Figure 4.6).

At this point, another consideration would be the amount of light each projector would deliver to the screen, at the

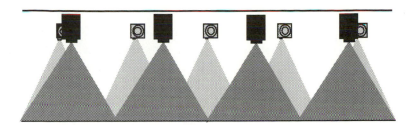

distance mentioned the 250 W Carousel would deliver 380 lux whereas the BP2 would deliver 1721 lux, a very good figure. Finally the cost of hiring the two systems would have to be compared, and the cost of producing the slides.

If front projection was considered, then it can be seen immediately that more stage depth is possible, even allowing for a cross-over behind the 'screen' for the cast, a stage depth of 8 m is possible this way. In order to avoid distortion, the projector would have to be mounted centrally, perhaps suspended from a lighting bar. In this position the BP2 could provide a 10 m wide picture on its own although this would only be at 250 lux, and this would be rather low if the slides were highly coloured or detailed, and the ambient level likely to be high. It would be possible to use two projectors this way, and match the join together, in this format the illuminance would increase to 1083 lux which is a much better figure. Some keystone distortion would result if the projectors were suspended, and this could be overcome either by making pre-distorted slides, or angling the 'screen' accordingly. A disadvantage is the restriction placed on the movement of scenery, and also of actors, adjacent to the 'screen' caused by their having to avoid the projection beam.

Some of these restrictions can be minimized by placing the projectors at the sides of the stage, which opens up the main acting area for scenery. The illuminance values are about the same as for the above option, but again pre-distorted slides would be required. In this position the projectors are once more accessible in case of problems (Figure 4.7).

Another factor is the nature of the screen material, back-projection material is generally more expensive than front, and of course it also absorbs more light, so the front projection situation offers a number of advantages; it would, however, be perfectly possible to project onto canvas or a plaster cyc, thus, if money is tight, it can be concentrated onto the projectors where it is most needed.

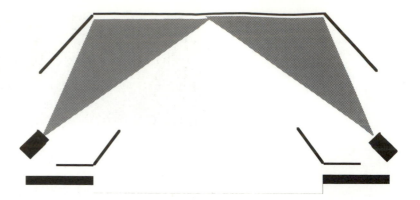

Figure 4.7 By moving the projectors into the wings the main stage area is freed for the movement of actors and scenery, but the slides will have to be pre-distorted to cope with the angle.

This case study has still other options, but at least the reader can hopefully see why it is important for set designers and directors to consult about the use and format of projection before too many decisions about the set (and screen location) are made.

Little Tramp

The author was asked to design the lighting for the UK première of the musical *Little Tramp* by David Pomeranz, about the life and times of Charles Chaplin, and the production required a constant panorama of scenes as Chaplin's life unfolded from London to Hollywood. Since many scenes were short, and since the theatre (in Basingstoke) had (then) a small stage, designer Elroy Ashmore opted for scenic projection. His design rested on a centre revolve which contained an arch construction, one side was christened 'posh' for the Hollywood scenes, and the other side was brickwork for the earlier scenes in Chaplin's life in London slums. The arch contained a back-projection screen which could be rolled up into the arch to permit the movement of people and furniture through it (Figure 4.8).

At first, the author considered using high powered machines with 180 mm slides, but the hire of these units was beyond the production budget. Since the production had gathered some momentum at the point this fact was appreciated, it was essential that the projections went ahead somehow, and so the use of gobos was considered. The dimension from the back wall to the screen was about 3 m but by specially selecting 2 kW CCT silhouettes with 40 degree beam angles, a satisfactorily wide picture could be achieved. One benefit of using the 'Sils'

Figure 4.8 Elroy Asmore's set
model for the world première of
Little Tramp Basingstoke England,
lighting by the author

Figure 4.9 The black box
represents the main position of the
revolve with three tiers of
silhouette backprojecting gobos; the
white box represents an alternative
position served by another battery
of silhouettes on the axis of that
position. Most gobos were changed
for others during the interval.

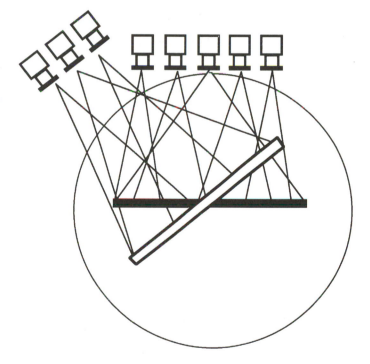

was that the revolve could be angled for some scenes, and yet
still receive an image from a Sil positioned on its new axis. This
would not have been possible with a single high powered slide
projector centrally placed (Figure 4.9).

In all, 19 Sils were used, although over 30 different projections were required. Additional Sils would not have been within the production budget to hire, and (perhaps!) more importantly, they could not have been accommodated within the required space around the centre line of the screen. Thus it was necessary to manually change the gobos over at the interval. With little facility to refocus all 19 spotlights during a complex interval change, it was hoped that the Act One focus would remain good for the Act Two gobos, and in fact this was the case. Each gobo holder had to be carefully labelled to which Sil it was to be inserted into for which Act. This production also contained several glass gobos which were used because the design required some elements to be suspended in light, and thus the gobos could not have metal links for support. The glass gobos were used with special mounts, and all the Sils were adjusted so that the light beam was 'flat', so that there was no hot spot to cause the glass to shatter. Nevertheless, one glass gobo did break because it was used in a modified holder, but fortunately spares had been purchased. The success of the projections was very much due to the crew's care in their use, and to the artwork produced by Elroy Ashmore, which wisely was clear and simple and lent itself well to suggesting the simpler days of movies.

Tobermory

The author was asked to light the première of an opera called *Tobermory* about a cat, which was chosen to open the new theatre at the Royal Academy of Music for which the author had been the consultant. The set designer required a front gauze, painted to resemble the outside of a house, to be set in place during the overture, and then as the overture merged into the first act, a dissolve through the gauze was to reveal the drawing room of the house behind. At the end of the opera the designer wanted a reverse dissolve to take place, by flying in a gauze slowly in front of a tightly lit part of the drawing room, and then dissolving back to the front of the gauze, which on this occasion, was to reveal an enormous head of the cat. It proved beyond the budget to purchase and paint two gauzes, and so scenic projection was considered. The only location for the projection was the lighting control room at the rear of the circle, and since this presented an angle square to the gauze there was severe danger of the projection beam passing straight through the gauze, and illuminating the scenery

Plate 3.1 Surtitle screen over the set, Herbert Senn and Helen Pond's design (and stunning scene painting) for *Don Pasquale*, Opera Company of Boston, lighting by the author

Plate 3.2 Jean Michell Jarre at Mont St Michel. Photo courtesy of Tony Gotelier

Plate 3.3 Panis at work in Vienna providing advertising over 3600 square metres. Photo courtesy of Ludwig Pani

Plate 4.1 Timothy O'Brien and Tazeena Firth's bold set and projection design for *The Knot Garden* at the Royal Opera House, 1970, lighting by John Bury

Plate 4.2 Gobos streak across the night sky in Simon Ash's set for the world première of *Winnie the Pooh*, lighting by the author

Plate 4.3 Simon Ash's evocation of the Shepard drawings for Vanessa Ford's world première production of *Winnie the Pooh*, directed by Richard Williams, lighting by the author, this scene shows the stage floor dappling effect with gobos from 20 × 2k profiles

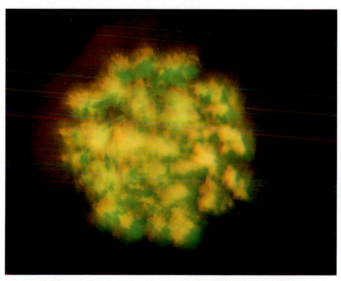

Plate 4.4 DHA gobo # 405 with split colour filters and projected out of focus. Photo courtesy of DHA Lighting Ltd

Plate 4.5 Composite back projection for *The Trials of Sancho Panza*, designed by Tony Walton, lighting and projection by Richard Pilbrow. Photo courtesy of Richard Pilbrow

Plate 4.6 Front projection by the author of the designer's artwork onto sharkstooth gauze, the set behind the gauze can just be seen top right

Plate 4.7 One over the Eight, London, 1961, set by Tony Walton, lighting by Richard Pilbrow, projections by Richard Pilbrow and Robert Ornbo. Photo courtesy of Richard Pilbrow

Plate 4.8 Close up of the complex rear projection screen system used on *I and Albert.* Photo courtesy of Richard Pilbrow

Plate 4.9 Innovative rear projection from massed Kodak Carousels by Robert Ornbo who also lit the production *I and Albert*, London, 1974, set by Luciana Arrighi. Photo courtesy of Richard Pilbrow

Plate 4.10 Patrick Robertson's front projections for *Boots with Strawberry Jam*, Nottingham Playhouse

Plate 4.11 Projection onto a white velour curtain by Andrea Arden Penn for a nightclub entertainer popular in Acapulco and Mexico City. Photo courtesy of Andrea Arden Penn

Plate 5.1 DHA animation wheels slash rain across a sharkstooth gauze in *Winnie the Pooh* directed by Richard Williams, lighting by the author

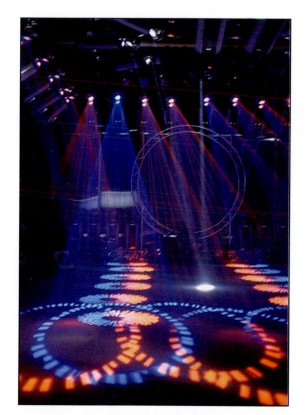

Plate 5.2 A light from Clay Paky's range of GoldenScan and SuperScan intelligent lights, controlled here by Pulsar's Oska. Photo courtesy of Lighting and Sound International

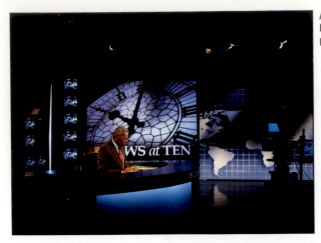

Plate 5.3 The Barcodata 5000 projects the face of Big Ben on London's 10 pm ITN Television News programme. Photo courtesy of WBN International

Plate 5.4 The Barcodata 5000 projects Cliff Richard high above the stadium on his *Access All Areas* tour. Photo courtesy of WBN International

Plate 5.5 High quality video projection from the Gretag Eidophor bringing all delegates closer to the speakers. Photo courtesy of Gretag

Plate 5.6 3D cone projected into smoke by Laser Innovations. Photo courtesy of Laser Innovations

behind before it was required. Tests proved that although much light did pass straight through, sufficient did return from the gauze itself to present a reasonable image and obscure the set behind, and so the decision to project was taken. The overall cost of the hire of the projector (a Reiche and Vogel 5 kW) and of making the opening and closing slide, was less than the cost of purchasing and painting a gauze, and the image of the cat's head appearing – seemingly out of nowhere (exactly like the Cheshire Cat) – and filling the proscenium arch was memorable (Plate 4.6).

Robert Ornbo

As one of the founders of Theatre Projects, Robert Ornbo has certainly made a considerable contribution to the theatre scene, but whilst he is perhaps best known today for his lighting of military spectaculars at London's Royal Tournament and the Edinburgh Tattoo, earlier in his career it was his experimentation with projection which was especially exciting. One his first productions was a revue called *One over the Eight* produced in London in 1961, and lit by the genius creator of Theatre Projects, Richard Pilbrow. Mr Pilbrow was clear that the types of large-scale projection which he had seen in Germany could be realized on the small stages common in the UK, and set about proving the fact. This was a considerable risk, given the little that was known in the UK about this technique at that time, and especially so, since the production in question had no other scenery! Two Reiche and Vogel 5 kw machines were used, each mounted on a tower behind a false proscenium, and focused to hit the cyclorama on the opposite side of the stage, a format which then became common. The slides were produced in conjunction with Robert Ornbo, who had to calculate all the distortion involved in the side projection position, but who could not undertake full-scale experiments as the Germans were then doing, because the revue was on tour and thus no full-scale space was available in which to work. The resulting pictures were 50ft wide and 25ft high and Mr Pilbrow writes, in his book, that the pictures pleased the critics so much that they praised the 'multitude of brightly-coloured backcloths'. Cleverly, the projection team (which included designer Tony Walton from whose work the slides were photographed) made a feature of the centre join, rather than trying to hide it and fail. However, within a short time (1966) Robert Ornbo was using the same split screen

Figure 4.10 *One over the Eight,*
London, 1961, set by Tony Walton,
lighting by Richard Pilbrow,
projections by Richard Pilbrow and
Robert Ornbo

technique, but with indistinguishable joins between the two
halves of the picture (Figure 4.10, Plate 4.7).

Robert Ornbo's projections in the musical *I and Albert*,
produced in London in 1974, introduced multiscreen projection into the British theatre scene. Luciana Arrighi's set
consisted of several levels, behind which were several rear
projection screens, shaped to resemble picture frames, behind
them was a large rear projection screen, into which apertures
had been cut, the same shape and size of the 'picture frames'
in front. The 'picture frames' could thus contain a galaxy of
famous Victorians, whilst the large screen was washed with
wallpaper. For other scenes both small and large screens could
be washed with the same image, thus creating a unified
backdrop (Plate 4.8).

Since there were so many slides, the Kodak Carousel was
used (fitted with a special 1200 w lightsource) and in many
ways this was a seminal use of the multiscreen projection in
the UK theatre since it demonstrated the Kodak Carousel's
ability to change slides quickly and reliably. One drawback was
the noise of the Kodak's fan, amplified by the 38 machines in
use, and thus a wall had to be constructed to contain this noise,
and fitted with conventional projection ports. Like so many
other significant projection shows in the UK at the time, this
one was the work of Robert Ornbo who also lit the show
(Plate 4.9).

Figure 4.11 Patrick Robertson's
projections for *The Cunning Little
Vixen*, Glyndebourne

Figure 4.11 Patrick Robertson's projections for *The Cunning Little Vixen*, Glyndebourne

Patrick Robertson

This quiet, self-effacing designer, has, with his partner Rosemary Vercoe, produced some of the most effective examples of scenic projection the UK has seen in modern times. Robertson's forte is to utilize some very simple idea as the core of his design, for example in *Boots with Strawberry Jam* for Nottingham Playhouse each screen was accompanied by a mirror at right angles to it, which thus reflected whatever was projected onto the screen. (This concertina format of the set was also used for *Elvis* in London which also involved front, back and cine projection.) For *The Cunning Little Vixen* at Glyndebourne in England, Robertson projected normal-sized trees when humans were on stage, but enormous leaves when the animals entered, all from a gondola suspended at the centre of the theatre's curved cyc and with some units back-projecting onto gauze covering the opening of the cyc, giving double images and enormous depth and mystery (Plate 4.10 and Figure 4.11).

The Hunting of the Snark

This so-called 'mega' musical, briefly inhabited London's West End from its opening in 1991, by all accounts it should have stayed longer. It will, however, be remembered for the largest (then) ever installation of slide projectors, a total of 96 Kodak 2060 Carousels. Many of the slides were hand-made by the show's creator, Mike Batt, who drew on top of theatre colour filter rather than relying on photographic film. This enabled him to be more flexible and make adjustments on site. The whole installation was the responsibility of Imagination, the London based design house, and their own rostrum-camera enabled them to produce composite slides which fitted perfectly together when projected side by side. The basic front projection module was 12 × 8ft and each was soft-edged to form pictures sometimes 40ft wide by 30ft high.

One advantage of having so many projectors, and of having a unit so reliable and responsive as the Carousel, is that it offers the possibility of animation by fast switching from one group of projectors to another. The *Snark* is not the first production to use this technique. For example, in *They're Playing My Song* at the Shaftesbury Theatre in London the stage was filled with animated dancers for a disco scene, front projected from the dress circle by a battery of Carousels. For the *Snark*, Imagination had to plan carefully in order to make the large number of projectors inconspicuous, and yet permit them to have a good throw to the screens. The programming was carried out at Imagination's studio on a third-scale model and it took six weeks in all. Control was by Genisis computers specially built by AVL of the USA, (Figures 4.12, 4.13 and 4.14).

Harvey Goldsmith's *The Planets*

In 1993 this familiar piece of music was given the spectacular treatment by the impresario Harvey Goldsmith, who had earlier brought opera within the reach of thousands through his presentations in arenas such as London's Earls Court.

The Holst suite was interpreted visually by stunning projections behind the orchestra, but of note was the development of scrollers carrying large format rolls of special plastic film more common on older (and smaller!) filmstrip projectors used in lecture work; two scrollers were placed in each projector (Hardware for Xenon 2ks) thus offering some overlapping images, and the use of the precision positioning, possible through the ETC AudioVisuel scrollers, offered speeds of change that varied from millimetre per minute to the almost instantaneous.

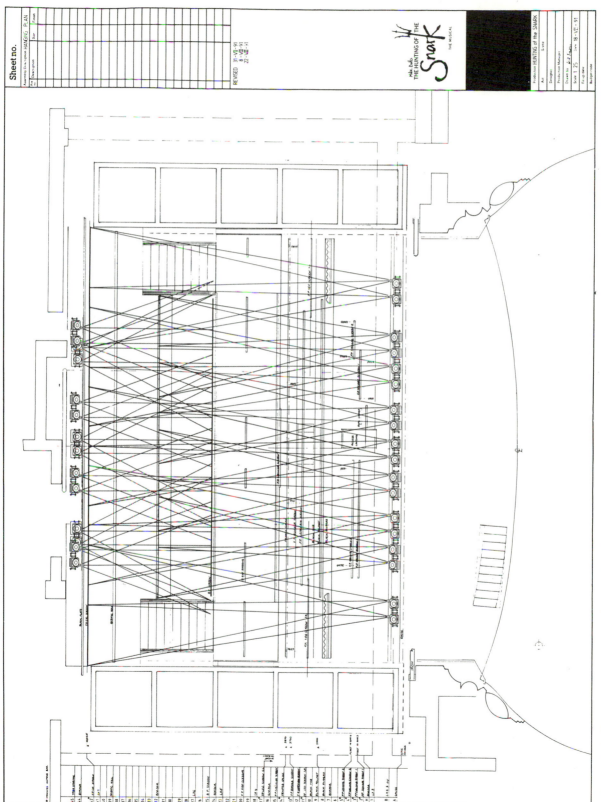

Figure 4.12 Plan of the stage projection positions for *The Hunting of the Snark*

88

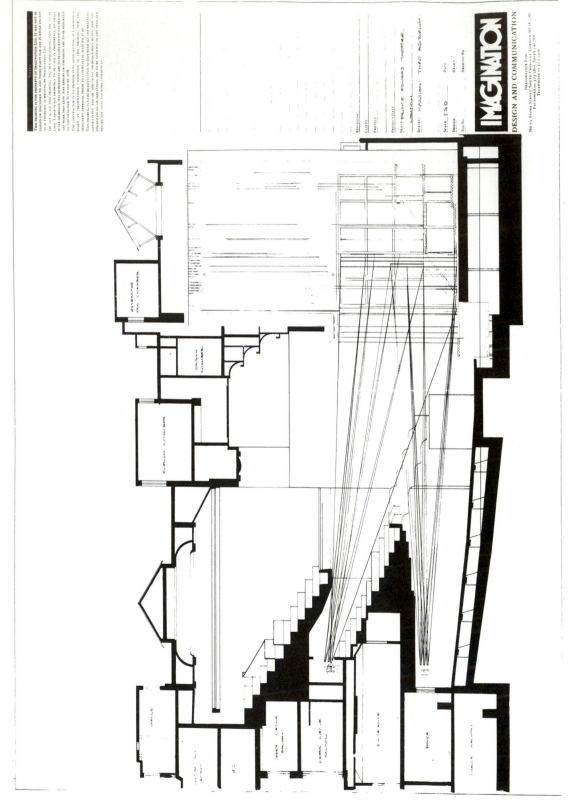

Figure 4.13 Section showing the FOH projection positions for *The Hunting of the Snark*

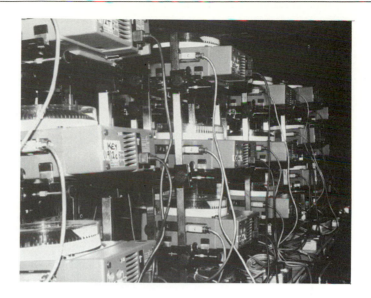

Figure 4.14 Part of the bank of projectors for *The Hunting of the Snark*

Leni Schwendinger

Leni Schwendinger is a rare designer, unafraid of projection, possibly because of her early experiences at the London Film School in 1973. Since that time her career has moved through lighting and set design and has been particularly inspired by her experience as a special effects controller at the Bayreuth Festival in 1979. Now also an architectural lighting designer, it is in the field of public art that she feels her work is most innovative. She has worked extensively around the world, notably in Japan where during *The Urban Heart, A Homebody?* (Tokyo 1993) she directed the audience members to cast giant shadows of their bodies onto the curved exterior of a biomorphic, cast concrete house. Her work also includes projections at Denver Airport in a shuttle train tunnel, images of letters and stamps on the columns of Manhattan Main Post Office, and paintings projected into the entrance to the Holland Tunnel in New York, (Figure 4.15).

Andrea Arden Penn

Andrea Arden Penn's work in projection has been seen in concerts by such artists as Barry Manilow, Boz Scaggs, the Captain and Tenille and Christopher Cross. Many of her slides for these concerts were hand-painted (sometimes also by light-

Figure 4.15 A typical example of Leni Schwendinger's work outdoors, in this case in Japan

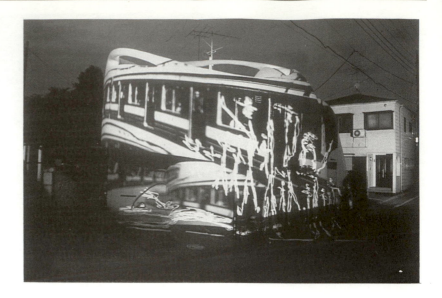

ing designers on the shows such as Bill McManus – and sometimes in conjunction with gobos selected by lighting designers such as James Moody). The advantages of projection in these situations are that it can be adjusted to fit almost any stage, which conventional scenery cannot easily do. Many singers also like to illustrate each song with some particularly relevant text or montage of photographs. (Plate 4.11).

Summary

Good projection is only achieved through good planning, and this involves the proper provision of budget, personnel and time to realize the various processes. Respect for the laws of optics is essential, and this should inform early discussions with the set designer who might also be the designer of the projection artwork, but not necessarily be technically aware of the factors which have to be taken into account to realize their design. Scale models can help to work out locations for equipment, masking and picture sizes, although much can also be achieved through the use of Computer Aided Design programs such as that offered through ModelBox.

The case studies illustrate various examples of the use of scenic projection, demonstrating that whilst the potential for change of the essential components is limited, many productions become landmarks through the innovative use of the equipment. Inevitably the most effective productions are those with the simplest ideas.

5
Moving projections

Effects

First, it is necessary to define the word 'effect' in the context of this book. Projected effects are those which concern a moving device, which, realistically or impressionistically, results in the image of such elements as rain, snow, fire and water. In each case, the projected effect should not be left to do the work on its own, rain, fire and water will each require careful sound in support; snow can also be produced by dropping tiny pieces of (fireproofed!) paper through slits in a sheet, fire will produce smoke, perhaps also alarms and mechanically breaking scenery.

As with all types of projection, the moving image will not show up if the level of direct and ambient light on the projecting surface is already high and consequently the light level has to be taken into account and planned in advance.

Most of the moving effects involve a rotating glass wheel or filmstrip which pass between a condenser lens and an objective lens. The wheels and filmstrips can contain actual photographic images, if a realistic effect is required, or an artist's rendering if a less definable image is required. The filmstrip has the advantage that if disposable it can be easily overmarked on the spot, whereas the wheel is a proprietary device, and cannot be altered unless the production owns it, since these are expensive ownership is unlikely although not technically impossible. Perhaps the best filmstrip unit is that offered by Pani, their AS-100 Image Scroller carries a 203 mm wide by 30.3 m long film, which can be moved through the

Figure 5.1 Pani universal film drive
fitted to a BP1.2 projector. Photo
courtesy of Ludwig Pani.

gate at a speed of 0.025 mm per second up to 178 mm per
second. Thus virtually non-repeating images are now possible.
The precision of the device is such that a special controller is
required, although this can be installed on an Apple Mac and
synchronized with audio and video systems (Figures 5.1, 5.2,
5.3 and 5.4).

Additionally, a similar device known as an 'animation
wheel' is available, this is simply a large aluminium disc with
carefully designed patterns cut out of it, and which works in
conjunction with a gobo to produce the effect of moving
images, whilst in fact it is projecting a still image of the gobo
(for example a rain pattern, cloud, flame) which is being
interrupted by the animation wheel itself and it is this which
suggests the movement. This latter device has two advantages over the more traditional effects wheels, in that not
only are they considerably cheaper to buy, but they will also
fit most conventional theatre profiles (lekos in the USA),
whereas the effects wheels require a proper optical projection system. Since the optical effects projectors have a
limited (although improving) range of lenses, the animation

Figure 5.2 Pani's 30 m long image scroller and controller. Photo courtesy of Ludwig Pani.

Figure 5.3 Pani rotating FX disc fitted to BP4 HMI projector. Photo courtesy of Ludwig Pani.

Figure 5.4 Pani double (overlapping) FX discs fitted to BP4 HMI projector. Photo courtesy of Ludwig Pani.

discs also offer greater flexibility in achieving the desired size of picture from a specified throw. All of these devices can be hired (Figure 5.5).

The author has used the animation discs many times, the proprietary devices shown here are available in the UK through DHA Lighting, and the company are eager to find new problems to solve in this area. For example the author needed a very random moving sea effect for a production, and DHA modified a variable beam profile so that one animation wheel ran through the gate of the luminaire, and another ran in the colour runners as normal. The number of combinations created by the two wheels, their variations in speed and direction and by the variable focuses offered by the two lenses, was considerable, and the effect was very flexible – a vital factor as the production was toured around many different theatres, and could not always be located the same distance away from the surface at which it was aimed. In another production the author successfully projected rain from two DHA animation wheels, fitted into two 2 kW profile luminaires, each of which had a selection of blue filters. The gobos were slashes angled diagonally, and by experimenting with the speed (10 rpm was eventually selected) a real downpour resulted. This was essential as the production was the London West End première of the authorized version of *Winnie the Pooh*, and the rain was needed for the famous flood episode in which Pooh sails off in an umbrella to rescue Piglet. The rain was projected from the theatre dress circle, onto a white sharkstooth gauze, so that the various episodes could be fading in behind as cameos, under the continuing rain effect (Plate 5.1).

All these devices are powered by electric motors and most are now available with variable speed (and direction). It is wise to test the effect at ground level since it will be easier to make adjustments to the speed and direction here than up a ladder.

Most of the effects come without colour, flame and some water and psychedelic effects being the exceptions, so that the designer must remember to insert his or her own colour choice – this applies particularly to the cloud effects. It is possible with these to project the effect of the setting sun onto the bottom of the cloud, leaving the upper part of the cloud, say, a darker blue, by splitting the filters top and bottom of the colour frame. Even when a colour is provided, changing it can create interesting effects, for example substituting blues and greens for the reds and ambers provided with the flame effects can produce a useful (possibly fast) running water, but only if the effect is slightly out of focus. It should be noted that most of these effects will in fact be used slightly out of focus.

Another point worth noting is that all effects look better if they are used in quantity, so that two cloud wheels really do look better than one. In almost all cases the wheels, filmstrips or discs should not be focused at the same speed, and some variation in colour is also useful. This overlap and contrast gives more depth to the effect and, in the case of snow for example, offers the possibility of having the snow come from (at least) two slightly different directions which is more realistic.

Readers might be familiar with the small effects projectors often used to project psychedelic patterns in discos, but more realistic discs are available, and the author has used these successfully on stage, notably in a pantomime when projecting moving clouds onto a front gauze. The advantage is that the projectors are very small and thus easily concealed.

The effect of flames can also be projected by the flame flicker wheel, a motorized disc which fits in front of a limited range of theatre luminaires and simply interrupts the light beam in an irregular way. Since very little light actually escapes, this effect will not register unless the stage is fairly dark, and so its use tends to be limited to fireplaces and other small conflagrations. Again, with the red and ambers removed and blues and greens inserted, this effect serves as a useful water ripple projection. There is also a very popular water ripple effect which is known as the 'tubular ripple', because the effect is created by the light passing through a tube, the sides of which contain slits. The advantage of this device is that it can produce a wide strip of formalized sea effect from a very close distance, as with all others, two are better than one, and split or broken colours better than a plain single sheet.

It is worth noting that flame effects can also be created very simply by waving strips of canvas in front of a luminaire, and waves by bouncing light off a tray of water which is caused to move slightly. Esquire Jauchem, the US special effects designer mentioned earlier (see Pepper's Ghost) once created special 16 mm projected flames for *Medea* by filming burning paper and then matting the film so that the flames could be restricted to specific parts of the set.

One of the most famous projection effects sequences comes in *The Flying Dutchman*, when the ghost ship is required to appear, and in the UK it is unlikely that the sequence designed by Charles Bristow for English National Opera will ever be bettered as an example of how to use the equipment in a choreographed sequence.

As one of the first post-Second World War opera lighting designers in the UK, Mr Bristow had an almost seminal effect on the genre. Typical of his work is his brilliantly simple idea which formed the base of the *Flying Dutchman* projection sequence. To provide the base colour for the sea and sky, Mr Bristow backlit the cloth with floodlights but masked the beams so that the only area the groundrow could light was the sky, and the only area the floods could light was the sea, and the position of the maskers was synchronized so that they created the perfect horizon line. Thus the colour and intensity of the sea could be adjusted without affecting the sky, and vice versa. Waves and clouds were projected from 2 kW optical effects wheels, and supported by tubular ripples working onto front gauzes to provide added mystery (Figures 5.6 and 5.7).

At the time of this sequence zoom lenses for scene projectors were virtually unknown, and Mr Bristow had to make the ship increase in size by crossfading from one slide to another, each showing the ship slightly larger each time. Many years later the author was required to make a projected ship enlarge for the UK première of *The Voyage of the Dawn Treader* by C. S. Lewis and adapted by Glyn Robbins. In the opening sequence, the *Dawn Treader* ship is simply a painting on a bedroom wall but, suddenly, it comes to life and takes over the whole wall, the children in the bedroom fall into the water and are rescued by the ship which then sails off into adventure. The

Figure 5.6 Charles Bristow's
lighting and seminal projection
sequence for *The Flying Dutchman*

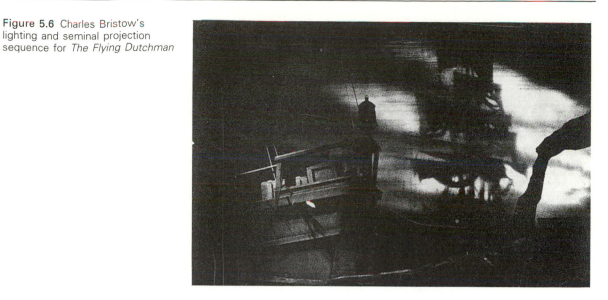

Figure 5.7 The foundation of
Charles Bristow's *Flying Dutchman*
in which a scenic groundrow and
border are positioned to mask
backlighting floods in such a way
that a horizon line is created and
each line of lights covers only part
of the backdrop. Effects projectors
were added over this.

bedroom wall was back projected, and the author's original
concept was to obtain a powered zoom lens and simply enlarge
the single slide at the appropriate time. However, since the
projector was very close to the screen there was no powered
zoom lens available for the wide angle required. Remembering
the Bristow *Flying Dutchman*, the author arranged for the
Dawn Treader to be photographed in increasing close-up, until

eventually its bows would fill the slide, and thus the whole bedroom wall. In all some 15 slides were used in the sequence, crossfaded between Kodak Carousels slung in a cradle behind the bedroom screen wall. One added bonus was that as each slide projected a slightly enlarged version of the ship, the centre of the ship shifted slightly, and it appeared as though the ship was rocking slightly on the waves, a happy accident! During the middle of the sequence, the stage lights had dimmed and smoke was poured in to conceal the removal of the bedroom furniture, finally as the ship was faded out, two real sections of it were trucked in from the sides, a really effective moment from the set designer Simon Ash.

Intelligent lights

There are a large number of spotlights which offer remote control over the basic functions of pan, tilt, focus and colour change, and which additionally can project a predetermined selection of (usually 6) gobos, the selection within the unit being made remotely. These devices are most at home in the lighting of pop music concerts where their lack of finesse (by comparison to that offered by pure theatre spotlights) is not a drawback. However, these words will become outdated during the lifetime of this book, as developments continue. Many intelligent lights can now fade rather than snap on or off, many offer a wide range of colours which are less saturated, and many operate more quietly than they did. Drawbacks are that the units do not compare with the more powerful theatre spotlights and thus would be unlikely to show up in large spaces where conventional theatre lights are already providing a basic area cover; kept to their own area, such as a projection screen this would not pose a problem however. Another drawback is their expense – some are only available on rental not purchase, at a minimum quantity and with their own operator. However, the ability to select a variety of gobos and colour, and move them at will is a stunning effect, especially when used with some sympathy for the movement and/or music of the production (Plate 5.2).

Film projection

The use of film in the live performing arts (at least in the UK) has been limited. It has been used in its own right where the scene on stage is set in a cinema, as in the famous sequence from *Evita* where the audience (both on stage and in the real auditorium) view her funeral. Elsewhere the technique has occasionally provided a moving background, sometimes for comedians in pantomime, who are supposedly driving a car or piloting a ship; alternatively it has been used to build up anticipation by showing an artist leaving home, driving to the theatre, arriving at the dressing room, and then walking onto the side of the stage before the artist's actual appearance in the flesh. Seldom has this writer seen a better marriage of film and live performance than that from the English team Chris and Tim Britton who perform as both Forkbeard and the Brittonioni Brothers. In their world, actors are (for example) filmed eating in a restaurant, when they can't pay the bill they run through a hole in the restaurant wall and emerge live in the theatre. Elsewhere characters appear at windows before coming into the room to hold conversations with themselves. This apparently simply idea is not damaged by the actors operating the projectors (16 mm) which are often clearly visible to the audience. On another occasion the writer recalls a production of *The Rocky Horror Show* in which the film of the musical was projected, and synchronized with, the action on stage at the appropriate point in the production.

Film projection – the equipment

The equipment involved in film projection is classified by the width of the actual film used (rather like slides) thus the common formats are 16 mm, 35 mm and 70 mm. As with slide projection, the wider the film, the more light which will pass through, and thus the potential for a larger picture is greater (Figure 5.8).

There are strict regulations governing the use of 35 mm and 70 mm projection equipment, although the regulations are sometimes relaxed for occasional use in specific circumstances, but the Local (licensing) Authority should be consulted at all times . Historically, this is because all the old film stock was highly inflammable; even today some xenon lamps emit ozone which requires the equipment to be properly ventilated by ducting, as do the older carbon arcs. The ventilation system must be separate from that for the audience. Whilst the smaller machines can manage on a 15 amp supply, the largest require 3-phase with large currents, and thus the electrical

Figure 5.8 A selection of film
stock, Super 8 mm, 16 mm, 35 mm
and 70 mm, note the actual width
of the aperture is less than this in
each case. *Source*: MOMI, p. 57.

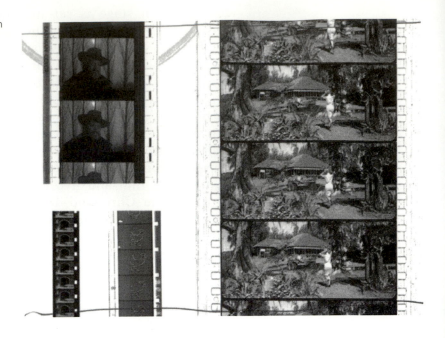

Figure 5.8 A selection of film
stock, Super 8 mm, 16 mm, 35 mm
and 70 mm, note the actual width
of the aperture is less than this in
each case. *Source*: MOMI, p. 57.

installations surrounding these systems have to be treated with respect.

Obviously the format chosen will provide the dimensions of the projectors required, and these dimensions will include the critical height from floor level to the lens. This is in turn informed by whether the projector is operating at a slight angle (usually this is the case) and therefore the projection port will need to be lower than the height of the lens to the floor. The projector will also require space around it for operation and maintenance, and since most projectors are operated from their right, looking at the screen, it is this space which is critical.

Projectors also require a wide range of supporting equipment such as rectifiers, rewind benches, amplifiers, changeover controls (from one projector to the other in dual installations) or towers (carrying large quantities of film) in single installations. In licensed premises, the projectionist must be in the room at all times, and so some space for this person must also be found. In many venues the position has to be shared with lighting and sound controls, and the most popular format in recent years is a long suite of rooms with a connecting corridor.

For these reasons 35 mm (and 70 mm) are unlikely to be used in live work, except where the venue already possesses a projection room, and/or where the production budget can accommodate the high costs of installation and shooting the film. Thus 16 mm is the more common format, and it has the advantages of being free of the regulations for larger formats, and having projectors which are smaller and therefore easier to build into the rig or the set. They are also self-threading, so they are easily operated by people with little or no previous experience, although whilst most offer automatic rewind, it is wise to do this manually so that the film can be inspected. Proper rewind benches can be made, so that this can be done whilst another reel is showing. Many 16 mm machines, which were designed for museum or display use (before the advent of video), can operate loops of film, and some can be remotely controlled.

Movie projectors use motors and their noise, coupled with the sound of the film moving through the various parts of the mechanism, really requires the machines to be removed from the auditorium, unless the production is producing so much sound itself that the machine's noise is masked; housings can be made to contain the sound, if a separate room is not possible – easy access and ventilation must be maintained.

The objective lenses for film can be changed to produce different picture sizes, and the format of the film also has a bearing. The various formats are identified through the ratio of the picture width to the height:

standard ratio	1.375:1
wide screen ratio	1.75:1
Cinemascope ratio	2.35:1
70 mm ratio	2.2:1

In some formats, (such as Cinemascope), the image is actually condensed on the film, and a special lens is required to open out the picture, these are known as Anamorphic lenses, and they are used in addition to the normal objective lens. The precise picture size is provided by the aperture plate in the projector's gate, which is filed on site to overcome minute variations in the throw. Note that the actual aperture of film is smaller than that of slides (in proportion), because of the amount of space taken up by sound tracks, so that the focal length of a particular lens would give different picture sizes if used for 35 mm slide and then 35 mm film, although few are actually interchangeable in this way.

In terms of the use of film in the live performing arts, it is assumed that each segment of film is unlikely to be very long, because the medium is essentially live rather than cinematic. The running time of film which can be contained on one spool varies with the format, so that a 600 mm spool of 16 mm film would last 55 minutes, but a 600 mm spool of 35 mm film would only last only 22 minutes. Hence the choice of format informs the size of spools which have to be used for a given length of running time, although most projectors would take 1800 mm spools, offering considerable increases in the above times. Commercial cinemas usually have two projectors so that one can take over from the other whilst the next spool is being loaded, or long reels of film held horizontally onto which entire programmes can be mounted – known as 'cake-stands' (Figure 5.9).

Commercial cinemas also have properly designed screens which are usually angled to face the projection room in order to overcome vertical (keystone) distortion, and curved so that the edges of the screen are the same distance from the lens as is the centre. The screens contain motorized black serge masking, which is automatically adjusted by the controller for the projectors, so that the amount of screen displayed matches the format of the film (the controller can also operate curtains

Figure 5.9 A large 35 mm cine projector. *Source*: MOMI, p. 57.

and house lights). Manual controls are also available. Professional cinema screens are also perforated, so that loudspeakers can be mounted behind them to ensure that the visual and aural pictures match.

In a live context, the screen could be a proper professional surface, or it might be part of the set. In both cases it should be noted that professional cinema screens are 'high gain' (see Chapter Two) and thus unless this approach is adopted on stage, then the picture will not be as bright as is perhaps expected by comparison. Additionally, we are accustomed to seeing films projected in a darkened auditorium, and this is unlikely to be the case on stage where the ambient light could also damage the picture.

Films are often back-projected, and this is a possibility on stage. However, unlike slides, the film cannot be reversed, and so a mirror must be used to bounce the projection beam and reverse it. This might be useful anyway if the throw is insufficient, although some 16 mm lenses are provided with in-built mirrors to reverse the image.

The proportions of the screen to the stage and the auditorium, are important in order to prevent the audience straining their necks and/or eyes to see the most extreme positions. For example, guidelines produced for cinemas have suggested that the angle from the eye level of the audience to the top of the screen should not exceed 30 degrees for 16 mm, and 35 degrees for 35 mm. In most theatres this is unlikely because the interruption caused by the orchestra pit, and/or apron stage, would push the screen further away from the front row than could be the case in cinemas. Equally, as has been indicated above, the viewing time is unlikely to be as long so fatigue is unlikely to build up. Nevertheless, the comfort of the audience must be taken into account at all times.

70 mm is reserved for the largest screens, and for the special 'wraparound' type of presentation often found in planetaria, and other domes and theme parks; since its use is limited in live performance, it is outside the scope of this book although the same principles apply.

Television projection – overview

Unlike other sources of projection, television exists in another medium, that is, it can be viewed and enjoyed without its being projected; this is not true of slides, or films. Thus, the application of projection to the television image, brings another dimension, in that it can be enjoyed by significantly more people than could cluster around a single television monitor.

Generally, but not exclusively, television projection is used to convey information, rather than to form one visual component of a whole design. Consequently, its use has tended to be restricted to conferences and trade launches, and the size of the budgets associated with these prestige events has enabled great advances to be made in the technology in recent years. Elsewhere, the larger machines have been substituted for film projectors in cinemas and arenas, and have enabled large audiences to view events, either happening across the world (such as live coverage of sports events) or happening in the arena itself, but at too great a distance to permit detail to be visible; recent examples illustrated here include Cliff Richard's *Access All Areas* tour and *Starlight Express* at London's Apollo Victoria Theatre (Plates 5.3 and 5.4).

Large-scale video projection, whilst still feasible, now has competition as video walls become more popular. These vast assemblies of television monitors, with picture edges almost flush, can be programmed to demonstrate a variety of images, large and small, and require somewhat less care about ambient light than would conventional projection. Detailed analysis of these systems is outside the scope of this book (since they do not involve projection) and the reader is directed to the bibliography, notably to Robert S. Simpson's *Effective Audio Visual*. In the arena, the television projection also has competition from screens composed of thousands of computer controlled coloured elements (typically but not exclusively LCDs with fluorescent tubes) and currently 'jumbo' screens up to 45 m wide have been constructed in this way (Plate 5.5). As these screens do not involve projection they are also outside the scope of this book.

Television projection – equipment

It may be appreciated that light is composed of the three primary colours, red, green and blue (RGB), and thus any other colour of light can be mixed by the relative proportions of these three colours. Video projectors tend to fall into two categories, dependent on how they use these three colours. One type of projector contains three Cathode Ray Tubes (CRTs), each of which receives the same picture, but is tuned in such a way that it only contains the red, or green, or blue component. Large diameter lenses are then mounted in front of the CRTs to collect the light, and focus it onto the screen. These systems generally do not offer much flexibility of throw, and consequently of picture size, because the lenses tend to be preset for a specific distance so that the picture is perfectly in focus. Many systems like this are mounted in housings which contain the screen in either front, or back-projected format. The latter is less satisfactory, since the light sources can sometimes be seen as a hot spot on the screen. Both front and rear projection types can suffer from restricted viewing angles, and thus they are most successful in small situations such as aircraft cabins (where they are now being replaced by armrest liquid crystal display screens), and also in hotel or domestic lounges and seminar rooms.

The other type of video projector uses more conventional projection light sources, such as the xenon, and consequently can achieve a brighter picture than the CRT-based systems described above. In the xenon systems, the television picture is produced by scanning a transparent surface which is covered with a fluid, usually oil. The scanning process produces an image on the transparent surface, which is then projected via the xenon lightsource and a lens system. The scanning system, and its attendant transparent surface, are contained within a vacuum valve and thus these systems are known as 'light valve projectors'. Small systems tend to have just a single valve for colour or monochrome and only require one lens, enabling them to be set up easily. In addition, they provide a variety of picture sizes from different throws. Larger machines use three RGB valves, and consequently they can produce larger pictures. More recently, video projectors have combined the two techniques described above, by replacing the CRT screens with RGB LCD screens, behind which is a powerful lightsource. These systems are smaller than the older light valve systems, and again are single lens which aids set-up time.

Most video projectors offer an electronic adjustment to the picture to overcome distortion, so that some variation is possible in the positioning of the unit with respect to the centre line of the screen; the adjustment includes reversal in the case of rear projection. Another facet of recent video projectors is the in-built decoder which enables the device to talk to the many standards of video that exist around the world. Finally remote control is also possible, a vital factor in stage presentations if the unit is in an inaccessible position, such as on the dress circle front or within the overhead lighting rig.

The future probably lies with the laser video projector, in which RGB laser beams are scanned and synchronized to form a single beam, which is then focused onto a rotating multifaceted mirror. This produces the 625 lines which currently make up the standard television picture. Since the laser is a coherent source, (that is, it does not diverge), then no lens is required, and the picture is always in focus.

Film and television projection – how bright?

The brightness of the picture in film and television projection, is governed by the same aspects as those for slide projection. Briefly, these are the intensity of the lightsource, the complexity and density of the image to be projected, the nature and reflectivity potential of the screen, and the amount of ambient light present.

16 mm film projection can be achieved with xenon arc lamps, comfortably providing a picture up to 10 m wide from the biggest machines, although the small portable variety provide about 700 to 2000 lumens, depending on the lightsource. The portable machines, therefore are unlikely to be suitable at more than 5 m wide, and that would be in ideal conditions. The 35 mm machines can be fitted with very large xenon lightsources, so that pictures up to 20 m wide are possible and common, although it should be noted that some smaller 35 mm cine machines are portable.

Television projection covers a similar range, in that the small CRT and single valve devices can produce 2000–3000 lumens, whilst the LCD devices can deliver 5000 lumens, quite sufficient for a screen 6–7 m wide, one such unit regularly provides the backgrounds for ITN's *News at Ten* on British television. The biggest machines, such as the famous Gretag Eidophor, can deliver 50 000 lumens which the manufacturers say is suitable for a picture up to 16 m × 12 m wide. It may be recalled that 1 lumen over 1 square metre is 1 lux. It is notable therefore, that, at the biggest picture size of 16 m, the Eidophor would be producing 260 lux, which would be quite acceptable, but care would need to be taken to prevent ambient light from washing out the picture; at 8 m wide the level would rise to over 1000 lux which would compete with virtually any stage lighting likely to be in the region. If this is compared with the smaller machines, such as those producing 5000 lumens over a 6 m × 4.5 m picture the level becomes 186 lux, which again

Figure 5.10 Currently the world's most powerful television projector, the Gretag Eidophor. Courtesy of Gretag.

would need care in terms of the ambient level and the reflectance value of the screen itself (Figure 5.10). (Note: the relationship of picture width to height must always be stated in these calculations.)

Where video is used to provide a large-scale blow-up of the live performer, then the lighting of the performer can influence the picture quality and intensity. Care has to be taken that the lighting so used does not add to the ambient, and wash out the projected picture, and in this it is useful to address the priorities of the event – which is the more important, the live action or the large picture?

The proliferation of video recorders and cameras, and the availability of small editing suites in High Street shops, makes VHS (video home system) much more accessible than film, so perhaps this means that it can find its way more easily into stage productions. This, however, is dependent upon the equal accessibility of the video projection systems. Certainly, if the delivery system is to a high standard, then VHS can produce good enough quality for stage work. It is, however, not as sensitive as film, and 16 mm projectors are not only more available than TV projectors, but they also work to an international standard unlike the many variations in the television industry; this might be important if the production has to cross national boundaries. It is also important to be aware of the differing quality of video and film, if the two are to be used in the same event.

Lasers

Lasers were once thought of only as sinister weapons, but today almost every cash register has some kind of laser to read the bar codes printed on the products. Other small lasers are used as pointers in school classrooms, and elsewhere they are used in almost every branch of industry. In terms of live performance, the laser beam is most associated with the pop and disco worlds and the vast resources of these areas has enabled great advances to be made over the years.

Today the laser beam is typically aimed at a scanning head, which contains shutters and mirrors to produce the many features required of live performances. A blanking shutter enables the beam to be interrupted, so that the line can be broken, this is known as 'dark move'; and RGB colour modulation provides a wide variety of colours. Mirrors controlled on two axes by galvanometers, enable the beam to be precisely directed to draw any shape. Standard movements and fonts are stored in the controllers, which can also record (typically 96) effects for subsequent replay (Figure 5.11 and Plate 5.6).

Lasers do attract the attention of safety officers, and users should be aware of the appropriate legislation. Briefly, the regulations require the units to be firmly mounted in such a position that the operator can clearly see the light paths, that they should be operated only by trained personnel, and that they should have cut-off switches adjacent. Lasers are classified by their power, and high powers are required to be used in such a way that the beam is 3 m above floor level (thus above head height – this may alter on stage if rostra are used). Other regulations govern the use of shutters, mirrors and setting up.

Figure 5.11 One of Laser
Innovations' drawing systems.
Photo courtesy of Laser
Innovations.

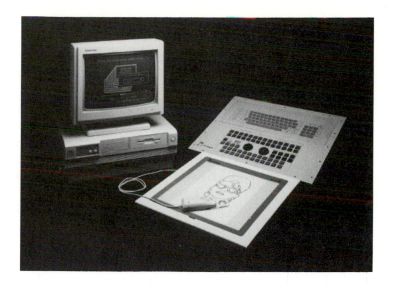

Although the effects produced by lasers are not as refined as would be possible by other methods, they carry a style of their own which could suit some productions.

3D projection

3D film projection was developed in the 1950s as one of the many techniques used to combat the threat of television, most of the others being concerned with widening the screen through Cinemascope, Cinerama and VistaVision. The audience was required to wear spectacles, one lens of which contained a green filter and the other a red filter. *Kiss Me Kate* (1953) and *Dial M for Murder* (1954) were both made in 3D, but released in 'normal' versions because the public disliked wearing the glasses. More recently in the UK, the BBC presented an evening of varied works, some of which were presented in 3D, but again the viewers were required to wear the special glasses, the abstract graphic effects on the television were perhaps more successful than the early attempts in the cinema, but it is hard to see how this system would be compatible with a live performance on stage which contained real 3-dimensional elements with which to compare (Figure 5.12).

Figure 5.12 A 1950s cinema audience wearing red and green spectacles for a 3D presentation, today spectacles are still needed but polarising filters can also be used as an alternative to the less sophisticated earlier 3D formats. *Source*: MOMI, p. 45.

An interesting variation has, more recently, been made by the Disney Company in their theme parks, in which 3D projection is achieved with polarizing filters in the spectacles so that the colour values remain accurate. The Disney screens are, like those in most commercial cinemas, perforated, so that smoke and laser beams can emerge from the screen at appropriate moments.

Projection has been used to help with 3 dimensionality on stage, the most famous example of which is probably the huge 'moving' head of Lord Olivier in the musical *Time* at London's Dominion Theatre. In this case, three films of Lord Olivier were made from different angles, and then projected onto the 'head' and synchronized so that it appeared that he was actually moving. This was a much talked about effect at the time, no doubt if the head was smaller then some degree of puppetry (borrowing from the sophisticated devices used on the Henson films and in *Spitting Image*) would be used. The author has, however, seen a similar device equally as effective (and considerably cheaper!) in a window display in Hull in the UK. In this situation a small (presumably 8 mm) film projector was providing movement up onto the face of a Father Christmas, the effect was excellent even from close to.

At the time of writing it is too early to assess the impact in New York of Sony's IMAX 3-D system which is about to open. The system utilizes one of the largest screens in the world with a steep bank of seating for good vision. Patrons wear a headset containing LCD lenses which blank out the eyes momentarily when triggered by computer controlled infra-red beams. Cameras for the format are expensive and at this time there is a severe shortage of product. However, it has been suggested that this system could form the basis of 3-D television in the future.

Summary

The projection of moving effects is achieved through the use of rotating discs or filmstrips, the speed and direction of which is usually variable. The effect should also be accompanied by sound, and the associated mechanical consequences of the effect involved. The ability to control remotely, and with high accuracy, the variations on pan, tilt, colour and beam angle, have produced a wide range of intelligent spotlights, most of which can also project a pre-determined selection of gobos (or patterns). These devices can be used to provide a number of fixed effects in different locations, or can be moved during the action of the production.

Film projection is rarely a major component of live productions, and large-scale film projection is expensive to achieve, partially because of the high cost of installing the protected projection room (usually) required by the licensing authorities. 16 mm projection is more affordable and versatile. Television projection is more associated with sports and conference events but the ability to utilize VHS material makes this more attractive than film. Some video projectors have a fixed throw to the screen but most offer more flexibility and the largest offer outstanding picture quality.

Laser and 3D projection are rare in the live performing arts, the former being more associated with disco work, and carry strict regulations beyond a certain power (however, recent improvements offer potential for drawing pictures); the latter requiring the application of coloured or polarizing spectacles. Some 3D moving effects can be created through simultaneous moving projections onto a still 3D object.

Moving effects, film, television, laser and 3D projection all require the same control over planning, location and ambient light required for still, scenic projection.

Projecting forward

Some years ago, Frederick Bentham, the life force behind the formative years of Strand Electric, and thus also behind the UK's twentieth-century stage lighting industry, remarked that light levels on stage have become progressively brighter over the years as spotlights have become more efficient and designers have used more of them. Whilst this may be true, the period in question also coincides with a refinement of the control of light so that greater precision is now possible in terms of focus, intensity and colour. Parallel with this, projection equipment has also undergone a revolution.

Not surprisingly the two main concerns of projection engineers and inventors over the centuries have been the development of brighter lightsources and the improved clarity of the image. With the advent of the 12 kW HMI lamp, certainly the first goal must have reached a plateau, although the lamp manufacturers are unlikely to rest on their laurels in this respect. The optical system has also reached hitherto unimaginable levels of sophistication through computer crafted compound lenses installed in computer designed optical systems (Figure 5.13).

Attention is now being given to improving the ancillary fittings which the projectors use, and the Pani Image Scroller, mentioned earlier, is a case in point. It is a long way from

Figure 5.13 The most powerful
projector in the world (at the time
of writing), Pani's 12 kW HMI BP12
Platinum. The company suggests
that this machine is so powerful
that it can be used for projecting in
daylight if the image is contained in
a small area! Elsewhere the BP12
can cover an area of 6000 m². The
projector uses a high performance
cooling system containing an air
conditioner and a turbo blower.
Photo courtesy of Ludwig Pani.

clockwork motors to the SMPTE time code or DMX512
controls which this device utilizes.

Elsewhere the LCD tablet, commonly associated with
overhead projectors, (both mentioned above) has inspired
some inventors to experiment with the concept of LCD gates
for spotlights. This would enable a variety of effects to be
delivered, beam shaping, dimming and, in the context of this
book, computer generated gobos. Perhaps within the lifetime
of this book, DHA and Great American Market (to name just
two) will be issuing their gobo catalogue, not on paper, but on
computer disk. Lightpens will enable the designer to shape his
or her own patterns and coloured LCDs will offer pictures
limited only by the designer's own imagination.

As with theatre spotlights, the future is likely to involve new
lightsources, cooler sources have long been known to point to
a number of benefits such as plastic housings, which would be
lighter to rig and easier to remote control. Cooler beams place
less demands on LCDs and variable focus Polythene/liquid
lenses.

Of course the laser will feature more prominently in future systems and undoubtedly they will continue to become more sophisticated so that their facilities take them closer to those of the laser television projector described earlier, whereby the complexity of the beams will become greater and the pictures less linear. To some extent this is already happening. The laser television projector itself offers perhaps the most scope for the incorporation of television projection into live performance since the image would remain clear irrespective of how the scenery or the performers moved, and of whether the image was a static projection or moving. Coupled with the ease (and economy) with which computer-generated images can now be created, the laser projection promises to bring a new dimension to the performing arts. Perhaps the frequency with which lasers (and other near-coherent sources) are now used brings the creation of three-dimensional images through the interaction of light closer.

All this technology already provides (in part) and promises (even more so) hitherto unimaginable images. In theory therefore our stages should be full of stunning projections, but these are few and far between. It is important that we have the people to use these new devices creatively, and incorporate them sympathetically into the production, without their very nature becoming distracting. It may be that scenic projection, *per se*, has been declining in the UK because of the dominance of three-dimensional design in the training of set designers.

Certainly in the UK, the set designer's art has reached exceptional heights and many designers have been rewarded internationally with awards, notably on Broadway and in the Prague Quadrennial. However, few seem to bring forth the kind of projections which Tim O'Brien, Robert Ornbo and Patrick Robertson did years ago and which still stay in the mind of those who saw them. This triumvirate, by no means the only people experimenting with projection, nevertheless pushed forward the boundaries of the art. Given the amount of unknown factors this must have taken considerable courage since in almost all the shows in which they used projection, there was little else had it failed. Perhaps this itself provides a clue to projection's current low profile. Art forms require the ability to fail in order that they might succeed and whilst much of the projection process can be calculated, good projection nevertheless requires the opportunity to experiment.

The demand for accountability in modern theatre, notably in the UK, has presented a doubled-edged sword to the creative team. On the one hand the restrictions on resources have been seen as a challenge, and many emerging designers are producing exciting work with very little. On the other hand, the element of risk and experimentation has declined, and it is in this territory that projection has thrived in the past.

For the future, the development of smaller and brighter projectors, now enables them to begin to compete with display lighting in leisure and commercial areas, themselves benefiting from a more sophisticated (almost theatrical) design input. Themed restaurants and hotel lounges could enjoy a projected vista slowly changing during the day as the mood and cuisine adjusts to the customers' taste. Supermarkets and department stores could project commercials around the walls to attract attention to the latest offer. Already some cars have data projected up onto the windscreen in the manner of fighter aircraft, and perhaps projection will form a greater part of information displays in other transport areas. In short projection is likely to become a greater part of our everyday lives.

This can surely only de-mystify the technique and stimulate more involvement in the performing arts. It is hoped that this book has also gone some way to de-mystify the art of projection and shown that, once the techniques have been treated with respect, projection can provide an exciting contribution to the performing arts which no other discipline can match.

Bibliography

Altick, Richard D., *The Shows of London*, Belknap Press, Cambridge MA, 1978.

Association of British Theatre Technicians, *Update, ABTT*, London (monthly publication).

Association of Lighting Designers, *Focus*, ALD, London (monthly publication).

Bentham, Fred, *Sixty Years of Light Work*, Strand Lighting, London, 1992.

Fitt, Brian and Thornley, Joe, *The Control of Light*, Focal Press, Oxford, 1992.

Freeman, Michael, *Cameras and Lenses*, Collins, 1990

Freeman, Michael, *The Photographer's Studio Manual*, Harper Collins, 1991

Ham, Roderick, *Theatre Planning*, (rev. edn),

Hedgecoe, John, *The Photographer's Handbook*, (rev. edn), Ebury Press, 1982.

Liesegang, Franz Paul, *Dates and Sources*, The Magic Lantern Society of Great Britain, London, 1986.

Longhurst, R.S., *Geometrical and Physical Optics* (3rd edn), Longman, 1973.

Nuckolls, James L., *Interior Lighting* (2nd edn), John Wiley & Sons, 1983.

Pilbrow, Richard, *Stage Lighting* (rev. edn), Nick Hern Books, London, 1979 and Drama Book Publishers, New York, 1991.

Rees, Terence, *Theatre Lighting in the Age of Gas*, The Society for Theatre Research, London, 1978.

Saxby, Graham, *Manual of Practical Holography*, Focal Press, Oxford, 1991.

Simpson, Robert S., *Effective Audio Visual: A User's Handbook* (2nd edn), Focal Press, Oxford, 1992.

Smith F.G, and Thomson J.H., *Optics* (2nd edn), John Wiley and Sons, 1988.

Strand Lighting, *Lights*, Strand Lighting London, (published several times annually)

Strand Lighting, *Tabs*, (actually published by Strand Electric then Rank Strand) London.

The Magic Lantern Society of Great Britain, *The New Magic Lantern Journal*, London (regular publication).

The Professional Lighting and Sound Association, *Lighting and Sound International*, Eastbourne (monthly publication).

Thompson, George (ed.), *The Focal Guide to Safety in Live Performance*, Focal Press, Oxford, 1993.

Thorn Lighting, *Technical Handbook*, published by Thorn Lighting and updated by them at regular intervals.

Turner, Adrian, *Museum of the Moving Image,*, Museum of the Moving Image, London, 1988

Patricia MacKay (ed.) *Theatre Crafts International*, Patricia MacKay, New York (published 10 times per year).

Viefhaus-Mildenberger, Marianne, *Film und Projektion auf der Buhne*, Verlag Lechte Emsdetten (Westf.) 1961.

Glossary

Aberrations

Spherical and chromatic aberrations are distortions produced in the image by the way the lightbeam is refracted by the lens system, they are commonly identified by unwanted patterns, colours and shapes. Imperfections in the lenses, lightsources and reflectors which spread light to the edge of the lens (instead of concentrating it in the centre), and the differences in the refractions of compound lenses are usually the cause.

Animation wheel

A motorized wheel containing specially designed slits and which fits into the colour runners of spotlights, producing a simulation of movement, typically used with gobos in profile spotlights.

Aperture

The opening within an optical system which controls the brightness of the final image

ASA

Abbreviation for American Standards Association and a measurement of the speed with which film reacts to light. The ASA range is linear in that a film of 100 ASA is half as fast as one of 200 ASA and a quarter as fast as one of 400 ASA and so on. There are other measurements of speed, see ISO and DIN.

Beam angle

The angle within which the useful amount of light emerging from a spotlight or projector is contained. In theatre terms

the angle at which the intensity of the beam falls to half
that measured at the centre of the beam on axis (½ peak
angle).

Bi-plane
Description of the filament in theatre spotlight bulbs in
which the strands of the filament are bunched together,
which produces a clearer image in gobo projection and also
enables more light to be collected by the reflector (see
monoplane).

CRT
Cathode Ray Tube, the tube on which a television picture is
produced and which forms part of television projection equipment (see also **light valve**).

Chromatrope
A projected effect produced by revolving two patterned wheels
in the projector. The device was invented in the nineteenth
century and a contemporary version can be obtained.

Colour temperature (see Kelvin)

Compound lens
An assembly of several lenses which is intended to operate as
a single lens. Compound lenses are complex and are usually
found in cameras rather than projection systems.

Condenser
A system of lenses which is contained within the housing of
the projector and which serves to evenly distribute the light
over the object to be projected.

CSI
Compact Source Iodide, a form of lightsource, which cannot
be dimmed.

Dark move
The process in a laser by which the beam is stopped by a
shutter in order that the lines drawn by the laser can be interrupted and not continuous.

Daylight film
A film which is intended to be used in daylight without the aid
of flash, the film is also suitable for use when tungsten or

fluorescent lighting is used if the relevant corrective filters are placed in front of the camera lens or the lightsource.

Depth of field
The distance between the points within which the image can be clearly focused by the lens. Long focal length lenses used for long distances tend to have short depths of field and short focal length lenses used for short distances tend to have long depths of field. Depth of field is important in projection especially when the projector is at an angle to the projection surface.

DIN
An abbreviation for Deutsche Industrie Normen, a wide range of standards, in this case one of which is used to measure the speed with which film reacts to light. The system is logarithmic and increases in steps of 3 units thus 24 DIN is twice as fast as 21 DIN which is itself twice as fast as 18 DIN. The DIN and ASA system correlate so that 24 DIN is 200 ASA and so on.

Doughnut
A mask inserted in the colour runners of profile spots in order to sharpen up the image of gobos, the mask contains a small hole which is adjusted according to the amount of light and clarity required. Some people know the doughtnut better as a 'butcher's stop'.

Enigma, Paling's
The name for the persistence of the eyes' vision which aids the projection of movement. Although developed in the nineteenth century the phenomenon was known at the time of Ptolemy.

ƒ value
The ratio of the focal length of a lens to its diameter, technically to the aperture which could be reduced by slide mountings for example.

Fast film
The speed with which the film reacts to light, measured in ASA, DIN or ISO units. A film which reacts quickly to small amounts of light will be grainier than one which reacts more slowly. This could be critical if the film is to be projected over a large area.

Flat

The name for the adjustment to the burning position of a light-source relative to its reflector so that the light is evenly spread throughout the beam (see also **peak**).

Focal length

The distance between the centre of the lens and a location on the axis of the lens from which a point source of light would produce parallel rays of light through the lens passing into infinity. The focal length is sometimes called the 'focal distance' and the point from which the said rays would emerge is known as the focal point, it is at this position which the slide or film would be located. Wide angle (i.e. short throw) lenses have short focal lengths and narrow angle (i.e. long throw) lenses have long focal lengths.

Gate

The position in an optical system at which the object should be placed if it is to be focused by the objective lens, typically the gobo slot or point of access for the slide or disc.

Ghost, Pepper's

The system of causing a reflection to appear on stage and thus suggest a ghostly apparition and which is named after its inventor Professor John Pepper.

Gobo

Strictly an object placed within the optical system of a projector (usually but not necessarily at the 'gate') in order to block out some light and produce a specified image. Gobos are usually metal but glass versions are also available, complex effects can be created using very simple stage spotlights. The US term is 'pattern'.

Halogen, Tungsten (see Tungsten halogen)

HMI

An efficient form of lightsource which strikes like an arc and which therefore cannot be dimmed by the conventional means of varying the current. HMI equipment can be dimmed by adding a mechanical shutter into the optical path. HMI sources can flicker because of their dependency on alternating current (AC)

Illuminance

See **lux**.

Image
The result of the projection process rather than the item within the projector from which the result is obtained. The *Oxford English Dictionary* defines image as an 'optical appearance. . . produced by light. . . from an object. . . refracted through a lens'.

ISO
An abbreviation for the International Standards Organisation and one which like DIN covers a wide range of applications, again like DIN and ASA in this context it is used to measure the speed with which film reacts to light. The ISO rating incorporates both the ASA and the DIN.

Kelvin
A temperature scale which indicates the colour of the light-source and which matches the Celsius (Centigrade) scale so that each degree Celsius is the equivalent of one degree Kelvin, with the addition of 273, so that 100 degrees Celsius is 373 degrees Kelvin (abbreviated to K). Industrial and entertainment lightsources which produce a colour close to daylight have high values and those which are used in more domestic situations have lower values (see colour temperature).

Keystone
A form of distortion caused by part of the screen or projection surface being closer to the projector than the rest. Typically this is the top giving rise to a distinctive wedge shape of image.

Lantern, Magic
A form of projection in which more than one projector is employed but in which the beam of each exactly overlaps at the screen so that by crossfading from one projector to another a series of developing or apparently moving images can be created. The objects were usually, but not exclusively slides. The Magic Lantern (and Laterna Magica) first grew to prominence in the seventeenth century.

Leko (see Profile)

Light valve
A television projection system which uses valves containing a scanned image which is then projected via a high intensity light source.

Limelight
A form of illumination created through the burning of a piece of quicklime. The limelight was developed in 1822 and was then used in surveying and the idea was acquired by the theatre industry which used it as a (then) powerful light-source in followspots (hence 'being in the limelight') and projectors.

Linnebach
The form of shadow projection named after Adolf Linnebach who became the Technical Director of the Munich Opera from 1923 to 1944.

Luminance
See Table 3.1 on page 41.

Lux
See Table 3.1 on page 41.

Matt screen (and semi-matt)
Matt screens will appear uniformly bright from whichever angle they are viewed, semi-matt screens concentrate the light towards the centre and this appears brighter but has a reduced viewing angle.

Mechnical dimmer
A device attached to a projector in order that its light output may be varied but which cannot be achieved through varying the current because of the nature of the lightsource. Typically a pair of grey glasses moved by stepper motors moved to replicate the intensity scale of thyristor dimmers.

Metallic screen
A screen which contains metallic paint in order to increase the amount of reflection, such screens can be very directional.

Monoplane
Description of the filament of spotlight bulbs in which the strands of the filament are all in line (see **bi-plane**).

Newton's Rings
The effect produced by light passing through two surfaces separated by a gap which is sympathetic to the wavelength of light.

Object

In a projection system the object is the item which is being projected, it would usually be a slide or film.

Objective

The objective lens is the last lens in the optical system and comes after the object in the light path. In most projectors it is the objective lens which can be changed for another to produce a different size of image at a given throw and it is the objective lens which is used to adjust the focus of the image.

Overhead Projector (OHP)

A projection system in which the transparency is projected typically behind a lecturer who can have constant access to the transparency to overmark it or point to it. The OHP has limited focusing abilities and consequent limited projection range but computer generated graphics and text can now be projected via Liquid Crystal Displays (LCDs) sitting on the OHP light table.

Pattern (see gobo)

Peak

The name for the adjustment to the burning position of a light-source relative to its reflector so that the light is more concentrated in the centre of the beam (see also **flat**).

Plano-convex (PC) lens

Description of the cross section of a common lens used in projection (and theatre spotlights) in which one side is flat and the other curved with the centre of the curve being away from the flat side. Lenses curved on both sides are known as bi-convex.

Profile (spotlights)

The generic name in the UK for theatre spotlights the optical system of which resembles that of a projector and which can therefore be used for the projection of gobos. The US term is 'Leko' or 'Ellipsoidal'.

PAT

Portable Appliance Test(er), a testing system of a device's electrical safety carried out with a special meter and which can be used as part of the maintenance required by UK safety legislation.

RCD
Residual Current Device, a sensor which cuts off the power automatically if a fault is detected in a circuit. There are strict guidelines as to how these devices should be used.

Screen
Literally the surface onto which the projection beam falls and therefore onto which the image is created, thus almost anything could be called a screen but the term is also used to refer to proprietory devices which reflect light in various ways, or which permit back projection (see **matt**).

SLR
An abbreviation for Single Lens Reflex and the 'collective' description for most professional 35 mm cameras. SLR cameras incorporate a mirror through which the subject is viewed.

Stop
A measure of adjustment to the aperture, generally 'stops' are applied to reduce the amount of light involved by adding special filters, increasing the shutter speed so that less light can pass through the aperture, or increasing the f value of the lens so that it will pass less light. This is not a problem usually associated with projection where more light is usually required, but with the process of photographing the artwork to make the slides.

Surtitles
The projection over or at the side of the proscenium of translations of the libretto in opera productions.

Tadpole
The name is associated in the UK with the device produced by DHA Lighting to adjust a gobo in the gate of a profile spotlight. (Companies in the USA supply devices which serve the same purpose.).

Tungsten halogen
Generic name given to light sources which contain a halogen gas under pressure and which rely on the ability of the tungsten particles of the filament to return to the filament during the burning process thus prolonging the life of the lamp.

Xenon
A lightsource which strikes like an arc and which cannot be dimmed in the conventional manner but has to use mechanical means. Xenon lightsources are smaller than HMIs and thus produce clearer images, they also work on direct current

Technical data

Equations (repeated here for easy reference)

$$\text{Desired Focal Length (F)} = \frac{\text{effective size of slide(O)} \times \text{distance from project or to screen – throw (D)}}{\text{picture size (B)+(O)}}$$

$$\begin{array}{l}\text{Illumination} \\ \text{intensity} \\ \text{(in lux or} \\ \text{lumens/sq m)}\end{array} = \frac{\text{output of projector in candelas/lumens}}{\text{(distance between projector and screen in metres)}^2}$$

$$\begin{array}{l}\text{Area of} \\ \text{screen in} \\ \text{sq metres}\end{array} = \frac{\text{output from projector (in lumens)}}{\text{target illumination intensity in lux}\ (= \text{lumens/sq m})}$$

1 lux = 0.0929 foot candles (fc)
1 foot candle = 10.76 lux
1 foot = 0.3048 metre
1 square foot = 0.093 square metre

$$\text{tangent of half beam angle} = \frac{\text{half beam diameter}}{\text{throw}}$$

$$\text{length of outer edge of beam} = \frac{\text{throw}}{\text{Cosine of half beam angle}}$$

Reflectance patterns for Harkness screens showing how the 'high gain' Perlux (and Spectral 2000 and Video) screens reflect more light on the axis (at 0) than the Matt White screen although the light falls off more steeply. Figures do show, however, that the light does not fall off less than the average overall level for matt white.

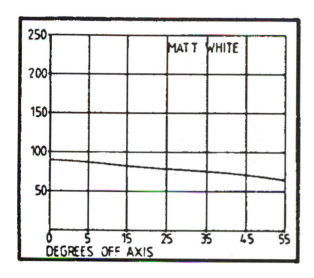

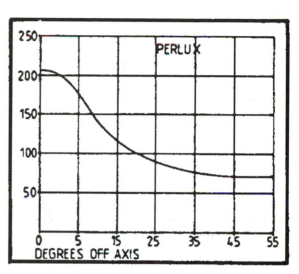

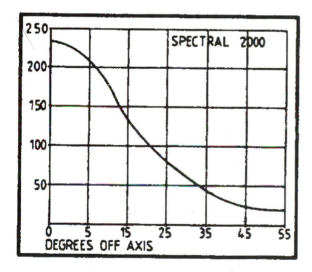

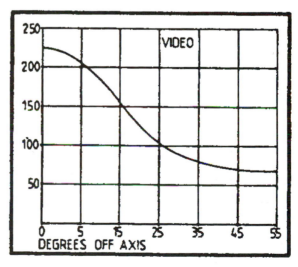

Data for Rosco back projection screens are given below.

ROSCO SCREEN

TYPE: BLACK (rear projection)

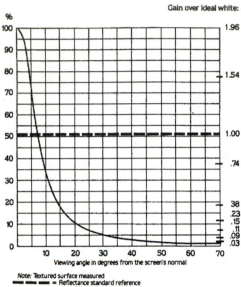

Note: Textured surface measured
━ ━ ━ = Reflectance standard reference

PCP Lighting Consultants Limited

ROSCO SCREEN

TYPE: TRANSPARENT (rear projection)

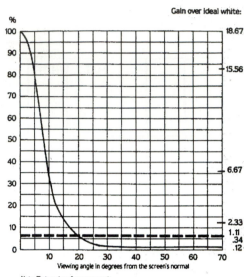

Note: Textured surface measured
━ ━ ━ = Reflectance standard reference

PCP Lighting Consultants Limited

ROSCO SCREEN

TYPE: WHITE (rear projection)

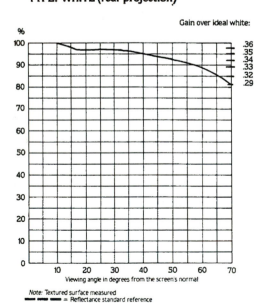

Note: Textured surface measured
━ ━ ━ = Reflectance standard reference

PCP Lighting Consultants Limited

ROSCO SCREEN

TYPE: GREY (rear projection)

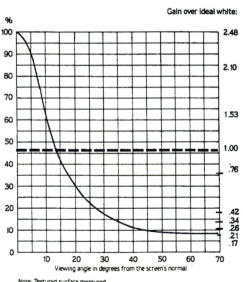

Note: Textured surface measured
━ ━ ━ = Reflectance standard reference

PCP Lighting Consultants Limited

The information contained in the figure below refers to the main projectors mentioned in this book as provided by the manufacturers (conventional theatre spotlights are excluded). Note that in some cases, such as the small 35 mm slide, overhead and small television machines there are other manufacturers.

Spreadsheet C1

manuf/agent	reference	basic details of system		
Kodak	Ektapro	24x36 & 40x 40 slides, 80-140 magazine, 300 w max lamp, remote /computer control		
		lenses (mm)26,36,45,51,55,60,85,93,100,136,150,180,200,253,70-120		
Hardware for Xenon	600 watt xenon	24x36 slides, dissolve, compact,		up to 7m wide
	1kw xenon	24x36 & 6x6cm slides, dissolve		up to 10m wide
	1.6kw xenon	24x36 & 6x6cm slides,		up to 14m wide
	5kw xenon	18x18cm slides, slide changer, dissolve		up to40m wide
	7kw xenon	18x18 cm slides, dissolve		up to 50m wide
		a booster is also available for Kodak machines		
Pani	BP 2/11	2kw, 18x18cm slides, discs and accessories		
	BP 5/11	5kw, 18x18 cm slides, discs and accessories		
	BP 1.2/HMI	1.2kw, 18x18cm slides, discs and accessories		
	BP 4/HMI	4kw HMI, 18x18cm slides, discs and accessories		
	BP 6/HMI	6kw HMI,18x 18cm slides discs and accessories		
	BP 12	12kw HMI, 18x18 cm slides, discs and accessories		
	BP2 blitz	as the BP2 but fitted xenon flash (lightning), note		
		other Pani machines have also been fitted this way		
		lenses (cm)11,13.5,18,22,27,33,40,50,60,80,125. prism, dimming shutters are available		
Great American	Mini Scene Machine			
		600w DYH, 1kw,2kw modular machine for 24x36,3",4"x5" slides and patterns		
		plus effects discs and rotators, 7' and 9.3" lenses		
	Scene Machine	1kw,1.2kw, and 2.5kw modular system for4"x5" slides (5 could be remote controlled)		
		plus discs, films, flicker wheels, prisms.		
		lenses (in)9,11,16,23.		
Strand Lighting	Cadenxa FX	2.5kw machine for motorised discs and 3.25" slides		
CCT	Starlette FX	2.5kw machine for motorised discs and 3.25" slides		
		lenses (in)2.5,3,4,5.5,6,8.		
Optikinetics	Solar 575	575w MH lamp 4000 lumen output machine, wide range of 6" FX wheels		
		lenses (mm)45,60 and 85-210 zoom		
Spotlight	Piccolo FX	1kw and 1.2kw modified spotlight for gobos and discs		
Barco	Barcodata 5000			
		LCD television projector, 5000 lumen output, suitable for screens 2m to 20m wide		
Gretag	Eidophor	light valve television projector with up to 50000 lumen output		
Eiki	2500M	Portable overhead Projector giving 4000 lumens output		

The figure below provides information about the image and throw of the Kodak Ektapro for its wide range of lenses.

Projected picture sizes

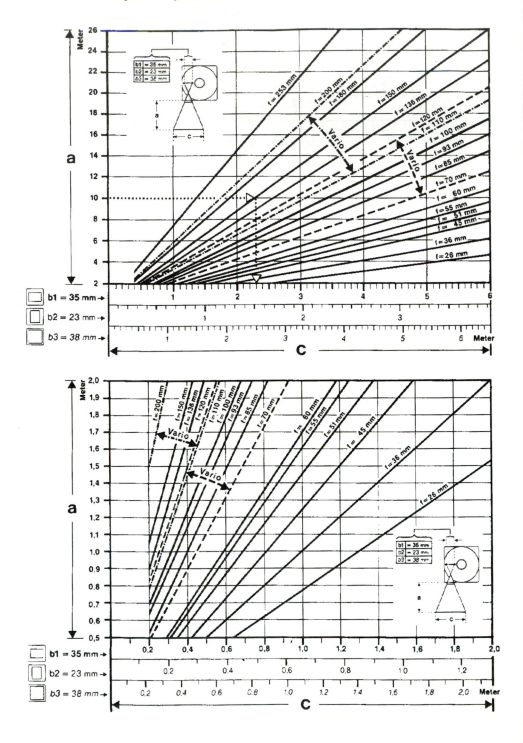

This figure is an example of information provided by manufacturers about the performance of their products, here the image and throw for Pani projectors when fitted with the effects discs. Generally the aperture for effects would be smaller than that for slides.

Projektionstabelle - Projection Table

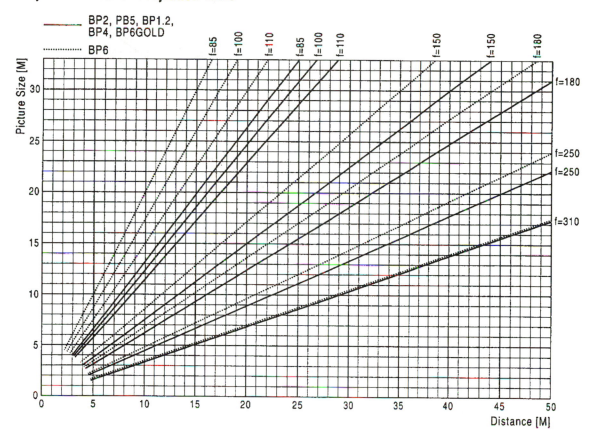

Effektobjektive - Effects Objetive Lenses

f = 80/100mm	G 951	f = 180mm	G 954
f = 110mm	G 952	f = 250mm	G 957
f = 150mm	G 953	f = 310mm	G 958

Vario-Objektive - Vario Lenses

f = 20 - 40cm	G 915
f = 30 - 60cm	G 916

Useful addresses

Association of British Theatre Technicians (ABTT),
47 Bermondsey Street, London SE1 3XT, UK
tel 0171-403-3778, fax 0171-378-6170

Barco Displays Division,
50 Suttons Park Avenue, Reading RG6 1AZ, UK
tel 01734-664611, fax 01734-267716

DHA Lighting,
3 Jonathan Street, London SE11 5NH, UK
tel 0171-582-3600, fax 0171-582-4779

Electrosonic,
Hawley Mill, Hawley Road, Dartford, Kent DA2 7SY, UK
tel 01322-222211, fax 01322-282282

Gerriets,
J 412 Tower Bridge Business Square, Drummond Road,
London SE16 4EF, UK
tel 0171-232-2262, fax 0171-237-4916 and
Road #1, 950 Hutchinson Road, Allentown NJ 8501, USA
tel 0609-758-9121, fax 0609-758-9596

The Great American Market,
826 N Cole Avenue, Hollywood CA 90038, USA
tel (213)461-0200, fax (213)461-4308

Gretag,
Althardstrasse 70, CH-8105 Regensdorf, Switzerland
tel 41-1-842-1111, fax 41-1-842-2100

Hardware SA -126,
Avenue Pablo Picasso, 92000 Nanterre La Defense, France
tel 47-76-00-29, fax 49-06-07-13

Harkness Screens,
The Gate Studios, Station Road, Boreham Wood, Herts WD6
1DQ, UK
tel 0181-953-3611, fax 0181-207-3657

Kodak,
PO Box 66, Hemel Hempstead, Herts HP1 1JU, UK
tel 01442-61122, fax 01442-240609

RoscoLab,
Blanchard Works, Kangley Bridge Road, Sydenham, London
SE26 5AQ, UK
tel 0181-659-2300, fax 0181-659-3153

Ludwig Pani,
Kandlgasse 23, A-1070 Vienna, Austria
tel 521-08-0, fax 526-42-87

Production Arts,
636 Eleventh Avenue, New York NY 10036, USA
tel (212) 489-0312, fax (212) 245-3723

United States Institute of Theatre Technology (USITT),
10 West 19th Street, Suite 5A, New York NY 10011-4206, USA
tel (212) 924-9088, fax (212) 924-9343

White Light,
57 Filmer Road, London SW6 7JF, UK
tel 0171-371-0806, fax 0171-731-3291

Photo acknowledgements

Simon Ash
Elroy Ashmore
Basingstoke Theatre Trust
Frederick Bentham
Lynton Black
Sarah Caldwell
Croydon College
James DeVeer
Elite
Wyatt Enever
Vanessa Ford Productions
Tony Gottelier
Great American Market
Gretag
Hardware for Xenon
Harkness Screens
Her Majesty's Stationery Office
DHA Lighting Ltd
Kodak
Laser Innovations
Paul Lawrence
Franz Paul Liesegang
Lighting and Sound International
Lightworks
M&M
The Magic Lantern Society of Great Britain
Modelbox
The Museum of the Moving Image

Timothy O'Brien
John Offord
Ludwig Pani
Andrea Arden Penn
Richard Pilbrow
Pitman
Helen Pond
Terence Rees
Patrick Robertson
The Royal Academy of Music
Herbert Senn
Leni Schwendinger
The Society for Theatre Research
Herman Sorger
Strand Lighting (Tabs)
WBN International
Richard Williams

and a special thank you to
Mervyn Heard

Index